SpringerBriefs in Food, Health, and Nutrition

More information about this series at http://www.springer.com/series/10203

Springer Briefs in Food, Health, and Nutrition present concise summaries of cutting edge research and practical applications across a wide range of topics related to the field of food science, including its impact and relationship to health and nutrition. Subjects include:

- Food chemistry, including analytical methods; ingredient functionality; physic-chemical aspects; thermodynamics
- Food microbiology, including food safety; fermentation; foodborne pathogens; detection methods
- Food process engineering, including unit operations; mass transfer; heating, chilling and freezing; thermal and non-thermal processing, new technologies
- Food physics, including material science; rheology, chewing/mastication
- Food policy
- And applications to:
  - Sensory Science
  - Packaging
  - Food Qualtiy
  - Product Development

We are especially interested in how these areas impact or are related to health and nutrition.

Featuring compact volumes of 50 to 125 pages, the series covers a range of content from professional to academic. Typical topics might include:

- A timely report of state-of-the art analytical techniques
- A bridge between new research results, as published in journal articles, and a contextual literature review
- A snapshot of a hot or emerging topic
- An in-depth case study
- A presentation of core concepts that students must understand in order to make independent contributions

Muhammad Riaz • Muhammad Zia-Ul-Haq
Bashar Saad

Anthocyanins and Human Health: Biomolecular and therapeutic aspects

Muhammad Riaz, Ph.D.
Shaheed Benazir Bhutto University
Sheringal, Pakistan

Muhammad Zia-Ul-Haq, Ph.D.
The Patent Office
Karachi, Pakistan

Bashar Saad, Ph.D.
Al-Qasemi Academic College
Baga Algharbiya, Israel

Arab American University Jenin
Jenin, Palestine

ISSN 2197-571X ISSN 2197-5728 (electronic)
SpringerBriefs in Food, Health, and Nutrition
ISBN 978-3-319-26454-7 ISBN 978-3-319-26456-1 (eBook)
DOI 10.1007/978-3-319-26456-1

Library of Congress Control Number: 2015957964

Printed on acid-free paper

This Springer imprint is published by Springer Nature
The registered company is Springer International Publishing AG Switzerland

Preface

During the last five decades, epidemiological studies as well as basic and clinical studies have consistently shown that there is a significant positive relationship between intake of herbs, fruits, and vegetables and reduced rate of chronic diseases in humans, such as cardiovascular diseases, diabetes, common cancers, and other degenerative diseases as well as aging. This is attributed to the fact that these products provide an optimal mix of basic macromolecules (carbohydrates, proteins, lipids, and nucleic acids) as well as dietary fiber, antioxidants, vitamins, and minerals. Various molecules in the diet can control the physiological functions of the body and supporting immune responses. Immune functions are indispensable for defending the body against attack by pathogens or cancer cells and thus play a pivotal role in the maintenance of health. However, the immune functions are disturbed by malnutrition, aging, physical and mental stress, or undesirable lifestyle. Therefore, the uptake of diets with immune-modulating activities is considered an efficient way to prevent immune functions from declining and reduce the risk of infection or cancer.

During the last years increasing consideration has been placed in plants and foods which can contain antioxidant substances. The chemical compounds present in plants that are related to health-promoting benefits considered several antioxidants, as vitamins C and E, carotenoids, and flavonoids. The chemical variety, molecular weight, three-dimensional conformation, and biochemical and physical properties of these flavonoids allow them to interact with different targets in many live organisms. Pigmented flavonoids, mainly anthocyanins, are considered the most important group of flavonoids in plants having more than 600 compounds identified in nature. Anthocyanins are water-soluble compounds that provide color to plant tissues (leaves, stems, roots, flowers, and fruits) ranging from red, purple, to blue according to the environmental pH and their structural composition.

Regarding the human consumption, the high intake of foods rich in anthocyanins offers potential health beneficial effects on various disorders associated with cancer, aging diseases, obesity, neurological diseases, inflammation, diabetes, as well as bacterial infections.

This book explores and introduces information available concerning the structure, composition, and abundance of anthocyanins in fruits and its bioavailability and biological activity related to health-promoting effects. This book includes nine chapters, embracing particularly historical aspects and present uses of traditional Arab-Islamic herbal medicine. Chapter 1 focuses on botanical medicines or herbal medicines. These therapies are still utilized as the primary form of medicine by about 80 % of the world's population. Over 80,000 species of plants are in use throughout the world. Usually, leaves, fruits, flowers, seeds, and roots are formulated into tablets or pills, teas, extracts, tinctures, ointments, or creams. Currently, about 25 % of the commonly used modern pharmaceutical drugs are of herbal origin or contain at least one herbal-derived active compound. Indeed, some are extracted from herbal crude extracts; others are chemically modified to produce a pharmaceutically active drug that agonists plant active molecule. The therapeutic effects of medicinal plants are generally labeled as antidiabetic, anti-inflammatory, laxative, carminative, demulcent, antiseptic, or antitussive. Chapter 2 provides a general introduction on anthocyanins including their general chemical structure, their therapeutic efficacy, and their safety. Chapter 3 focuses on the chemical structures of anthocyanins. It provides understanding of anthocyanins' chemistry including their occurrence in nature, e.g., plants and the recent discovered anthocyanins. Chapter 4 describes the utilization of anthocyanins as natural color in food and beverages. Chapter 5 is subdivided into ten subsections, which describe intake, metabolism, and secretions of anthocyanins in the human body. Chapter 6 provides knowledge about the biosynthetic pathways through which these compounds are synthesized in natural system. Chapters 7–9 describe the state-of-the-art knowledge of in vitro, in vivo, and clinical literature regarding the efficacy and safety of anthocyanins, including anti-inflammatory, antioxidant, antidiabetic, and anticancer effects and prevention and treatment of degenerative diseases.

Sheringal, Pakistan — Muhammad Riaz, Ph.D.
Karachi, Pakistan — Muhammad Zia-Ul-Haq, Ph.D.
Jenin, Palestine — Bashar Saad, Ph.D.

Abbreviations

AC	Anthocyanidins
ACN	Anthocyanin
ACNs	Anthocyanins
ACN-3-gly	Anthocyanins-3-glycoside
AMPK	Adenosine monophosphate protein kinase
AN-gluc	Anthocyanins glucuronide
BHA	Butylated hydroxyl anisole
BHT	Butylated hydroxyl toluene
CBG	Cytosolic B-glucosidase
CD	Cluster of differentiation
CH	Chalcone
C_{max}	Maximum concentration
CNS	Central nervous system
COMT	Catechol-*O*-methyltransferase
COX	Cyclooxygenase
Cy	Cyanidin
Cyd-3-glu	Cyanidin-3-glucoside
Cyd-3-rut	Cyanidin-3-rutinoside
Dp	Delphinidin
EMIQ	Enzymatically modified isoquercitrin
FC	Flavylium cation
FRAP	Ferric reducing antioxidant potential
Gal	Galactoside
Glu	Glucoside
HDL	High-density lipoprotein
ICAM	Intracellular adhesion molecule
IL	Interleukin
iNOS	Inducible nitric acid synthase
LDL	Low-density lipoprotein
LPH	Lactase phlorizin hydrolase
LPS	Lipopolysaccharide

MAPK	Mitogen-activated protein kinase
MCP	Monocyte chemotactic protein
MCP-1	Monocyte chemotactic protein-1
MDA	Malondialdehyde
MI	Myocardial infarctions
Mv	Malvidin
NO	Nitric oxide
ORAC	Oxygen radical absorbance capacity
PACNs	Pyranoanthocyanins
PB	Pseudo-base
PC	Protocatechuic acid
PDE	Phosphodiesterase
PEDF	Pigment epithelial-derived factor
Pg	Pelargonidin
Pn	Peonidin
PPAR γ	Peroxisome proliferator-activated receptor gamma
Pt	Petunidin
QB	Quinoidal base
RDI	Recommended daily intake
R_{max}	Maximal rate of excretion
RNS	Reactive nitrogen species
ROS	Reactive oxygen species
RPE	Retinal pigment epithelial
Rut	Rutinoside
SAOC	Serum antioxidant capacity
SGLT	Sodium-glucose co-transporter
SREBP1c	Sterol regulatory element-binding protein 1c
STZ	Streptozocin
SULT	Sulfotransferase
TAS	Total antioxidant status
TBARS	Thiobarbituric acid reactive substances
TEAC	Trolox equivalent antioxidant capacity
TNF-α	Tumor necrosis factor-α
TRAP	Total reactive antioxidant potentials
UDP-GT	UDP-glucuronosyltransferase
UV	Ultraviolet
VEGF	Vascular endothelial growth factor
VLA	Very late antigen
WHO	World Health Organization

Contents

Chapter 1
Diet and Herbal-Derived Medicines

1.1 Introduction

Health and food and their inter-relationship are one of most-debated topics by people of all age and income groups and are one of the most sought after and referenced topic in E-world. A peek at the magazine rack of nearby library or websites of newspapers confirms this as there are specific sections for healthy-eating there. Healthy-diet is affordable and does not include the side effects and the metabolic and physiologic burden that medication-packages impose on human body-systems. Diets rich in diversified eating pattern of plant-based foods are among the recommended lifestyle modifications to decrease the risk of diseases. Fruits, nuts, herbs, spices, grains and legumes and their products or by-products are an integral part of the cultural, socio-economic and health systems of all countries due to their established health-promoting effects and verified immunity-boosting claims.

Botanical medicines, or herbal medicines herbal medicines also known as phytotherapies, are the ancient healthcare remedies known to mankind. Hundreds of wild edible herbs and animal-derived preparations (e.g., milk, blood serum, urine, bones, and feathers) are utilized by traditional healers to prepare remedies for the treatment/prevention of all types of known illnesses as well as in maintaining healthy body, soul, and spirit. These therapies are still utilized as the main form of drugs by about 80 % of the world's population. Over 80,000 species of herbs are utilized for their medicinal properties throughout the world. Usually, leaves, fruit, flowers, seeds, and root are formulated into tablets or pills, teas, extracts, tinctures, ointments, or creams [1–4].

The last three decades have witnessed a tremendous growth in utilization of herbal-based diet and medicines as well as a significant progress in studying risks and benefits of these products at cellular and molecular levels. Herbs build an increasingly important source of new drugs. Currently, about 25 % of the commonly used modern pharmaceutical drugs are of herbal origin or contain at least one herbal-derived active compound. Indeed, some are extracted from herbal crude

M. Riaz et al., *Anthocyanins and Human Health*, SpringerBriefs in Food, Health, and Nutrition, DOI 10.1007/978-3-319-26456-1_1

extracts; others are chemically modified to produce a pharmaceutically active drug that agonists plant active molecule. The therapeutic effects of medicinal plants are generally labeled as anti-diabetic, anti-inflammatory, laxative (induces bowel movements or to loosen the stool), carminative (blocks the gas formation in the gastrointestinal tract or facilitates the expulsion of gas), demulcent (cover the mucous membrane with soothing film, healing pain and inflammation of the membrane), antiseptic (Healing from infections) or antitussive (cough suppressants). In contrast to pharmaceutical medicines, which are often synthetic and usually consist of a single compound), phytomedicines contain multiple constituents (Fig. 1.1) [4–6].

The popularity in utilization of the major traditional medicines has increased worldwide over the last half century. One of the main reason for the currently witnessed popularity is probably the belief that these medical systems have been used for hundreds of years and that natural product-based diet and herbal-based remedies are safe. The resurgence of interest in medicinal plant-based therapies at the global level has been so drastic that sales of these preparations in the world are estimated at more than 100 billion dollars per annum. Germany and France are the principal countries in Europe in the sale of plant-based preparation. The majority of German physicians (about 80 %) prescribe herb-based preparations [5–8].

Herbal medicines in the United States are sold as dietary supplements and in many European countries are classified as drugs, whereas in China and India as well

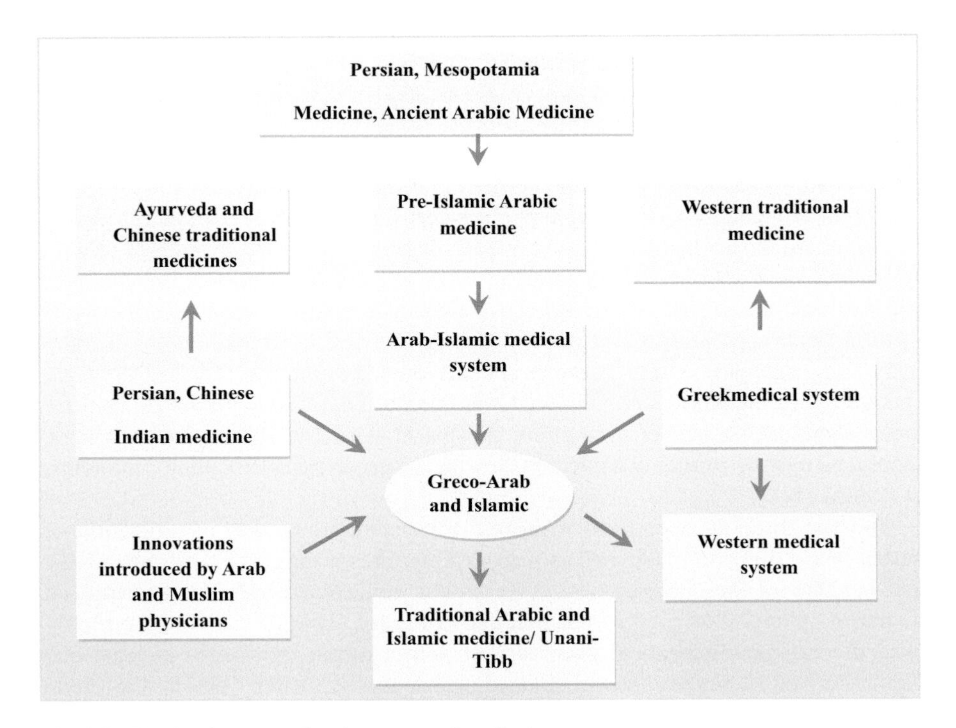

Fig. 1.1 The development of various types of medical systems

as in the Arab-Islamic world they are mostly sold over the counter without clear regulations. One of the problems with herbal drugs is that the concentrations of active compound(s) varies according to the soil, the weather conditions, and other environmental factors. Safety assessments of herbal products used in these traditional medicines have often been neglected due to their prolonged and apparently safe utilization. Nevertheless, a scientific evidence of the toxicity of such products has accumulated. This is not surprising, since herbal extracts consist of mixtures of tens of secondary metabolites, many of which are potentially toxic (e.g., hepatotoxicity, mutagenicity, and carcinogenicity). Therefore, the extensive consumption and fame of herbal products brought apprehensions and doubts over professionalism of healers and quality, efficacy, and safety of these products. Safety, contaminations, inappropriate preparation, or lack of knowledge regarding plant and drug interactions [6–9].

This chapter will provide a brief introduction to medicinal plants, including their therapeutic aspects and safety. In addition, the various major types of active compounds will be discussed. In the course of the following chapters, we intend to reveal the complexities, encourage comparisons various form of anthocyanins, and highlight relationship between their structure and functions. To keep within the scope of this introductory chapter, we will give a brief overview of the main topics of this book. The following chapters will comprehensively discuss the chemistry, metabolism, in vitro and in vivo scientific literature as well as the clinical significance of anthocyanins. We have organized this book around nine major topics, reflected by the titles of these chapters: (1) diet and herbal-derived medicines, (2) introduction to anthocyanins, (3) occurrence of anthocyanins in plants, (4) anthocyanins as natural color, (5) anthocyanins absorption and metabolism, (6) biosynthesis and stability of Anthocyanins, (7) the role of anthocyanins in health, (8) the role of anthocyanins in obesity and diabetes, and (9) anthocyanins effects on carcinogenesis, immune system and the central nervous system.

1.2 Current Status of Food and Herbal-Based Medicine

In parallel with the revival of interest in the traditional foods and medicinal plants, there is also an intensive research activity dealing with their safety and efficacy as well as with their action mechanisms at cell biological, biochemical and molecular biological levels. The modern clinical medicine is now beginning to accept the utilization of foods and herbal-based remedies once their efficacy and safety are scientifically investigated and validated. As a result, there is an increasing trend in Europe as well as in the USA and Canada to incorporate herbal-derived preparations as an obligatory course in the medical curriculum. Herbalists do not isolate a particular herbal active compound. In generally, they use the whole plant or water extracts from parts of plants, e.g., the leaves or roots (Fig. 1.2). They argue that the large number of active molecules present in herbs amplify therapeutic benefits and reduce possible side effects [6–13].

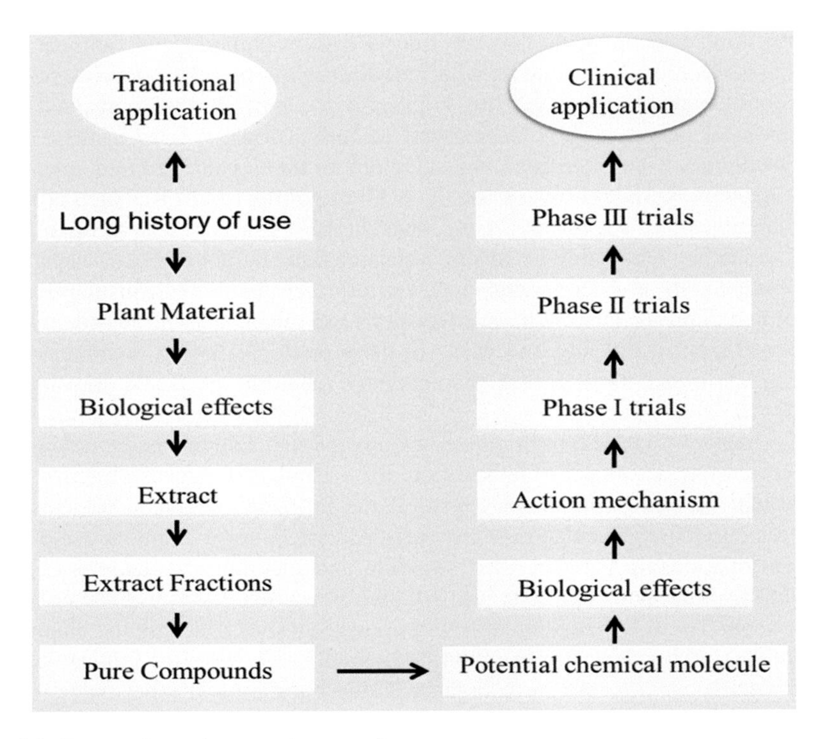

Fig. 1.2 Commonly used preparation procedure

Currently, products from biological sources, mainly from plants, animals, fungi, and algae represent a main source of new medicines. The World Health Organization (WHO) estimates that about 80 % of the world population presently utilizes plant-based preparations for prevention or treatment of all known diseases. Although financial funding for natural product–based drug development in the pharmaceutical industry has been reduced from 1984 to 2003, the percentage of natural product–based substances has remained relatively at relatively high level. Approximately 70 % of anticancer drugs developed since 1940s (155 drugs) have their origin from natural products [5–8, 13].

1.3 Medicinal Plants from Tradition to Evidence-Based Application

Medicinal plants are classified as wild-grown or as cultivated. As a result of natural selection, wild medicinal plants are grown in places with optimal soil and environmental conditions. These herbs may have been exposed to environmental pollution and pesticides. Hence, cultivated organic herbs grown under well controlled conditions represent a better alternative to wild-grown herbs. Organic vegetables and fruits are becoming increasingly available, as more and more farmers start to adapt the organic system. With careful management, organic food farms can provide high quality organic products to the market. In the case of medicinal plants, it is important that medicinal plant as well as fruits and vegetables have been harvested at the optimum season in order to achieve optimal therapeutic effects. For instance, many edible wild plants are collected in the spring, winter, and fall, but not in summer, when energy of plants is utilized for the purposes of flowering and growth.

To keep within the scope of this chapter we will give a brief overview of the most commonly used medicinal herbs in the Mediterranean region where more than 2600 plant species are found and about 450–700 plants are mentioned in various scripts for their uses as medicinal herbs. Plant parts used include seeds, roots, flowers, leaves, stems and fruits [1–5, 12–14].

1.3.1 Nigella sativa

The seeds of *Nigella sativa* (black seeds) are commonly used medicinal herbs throughout the Arab counties. These seeds have been used since dawn of civilization as a spice and food preservative and for the treatment and prevention of a wide range of diseases. Thymoquinone, dithymquinone, thymohydroquinone, and thymol are the main bioactive constituents responsible for the pharmacological properties of black seeds [14, 15]. Pharmacological and toxicological effects of the seeds have been extensively studied. In recent years, scientific journals have published numerous articles pointing to potential therapeutic properties *Nigella sativa*, including antimicrobial, hypotensive, antidiabetic, anticancer, antihistaminic, immunomodulatory and anti-fertility effects [8, 14–17].

1.3.2 Olea europaea

The olive, like *Nigella sativa*, is one of the most commonly utilized medicinal herbs throughout the Mediterranean. Both, olive oil and olive leaf are well known for their health benefits and have been for these properties for thousands of years. The primary medical constituents of olive leaf are the antioxidants oleuropein, hydroxytyrosol, hydroxytyrosol acetate, and the flavonoids luteolin, and luteolin-glucosides. Oleuropein and its hydrolysis products are those of the greatest therapeutic potential. Oleuropein has a vasodilator effect, increases blood flow in the coronary arteries and improves arrhythmia. It has proven to be a potent anti-inflammatory and antioxidant compound. Various scientific reports show that oleuropein also exhibits in antimicrobial activity against viruses, retroviruses, bacteria, yeasts, fungus, molds and other parasites [8, 14, 18, 19].

1.3.3 *Punica granatum*

The pomegranate has been used since long in Arab and Islamic traditional medicine to treat a wide range of diseases. These include anticancer, anti-inflammatory and anti-rheumatic activities. In Ayurvedic medicine the fruits as well as other parts of the plant are regarded as "remedy for every disease". For example, the bark and roots are used to manage vermifuge and helminthic conditions and the juice is considered as "refrigerant" and "blood tonic". Pomegranate flowers serve as a remedy for diabetes mellitus. The hydrolyzable tannins known as punicalagins show anti-oxidant effects in in vitro experiments. A Medline search using therapeutic potential and toxicological effects *Punica granatum* discloses greater than 350 citations, including antioxidant, cardiovascular protection, oral hygiene, as well as anti-inflammatory properties [8, 14, 20].

1.3.4 *Trigonella foenumgraecum*

Locally known as fenugreek, it is one of the most used herb in the Mediterranean. Most of the medicinal benefits of *Trigonella foenum-graecum* are found in the seeds, which have been used for thousands of years in Chinese and Ayurvedic medical systems. The anti-diabetic effects of fenugreek are mediated through the active compound, 4-hydroxy isoleucine. The action mechanisms include translocation of glucose transporter-4 (GLUT4) to the plasma membrane, delay of gastric emptying, slowing glucose absorption, and transport from the fiber content, as well as increased erythrocyte insulin receptors and modulation of peripheral glucose utilization [8, 14, 21, 22].

1.3.5 *Salvia officinalis*

Commonly known as sage, pharmacological and toxicological properties of *Salvia officinalis* leaf have been extensively investigated using in vitro and in vivo tests. The main active compounds of sage are present in its essential oil, which contains thujone, borneol, and cineole. In addition, sage leaf contains ursonic acid, tannic acid, ursolic acid, oleic acid, chlorogenic acid, fumaric acid, cornsolic acid, caffeic acid, niacin, nicotinamide, flavonoid glycosides, and estrogenic compounds. *Salvia officinalis* exhibits a wide spectrum of pharmacological effects. These includes antioxidant, anti-inflammatory, antimicrobial, carminative, weakly spasmolytic, astringent, and antihidrotic (inhibits perspiration). *Salvia officinalis* is also a well-known tonic and stimulant for the nervous and digestive systems. Clinical studies have demonstrated beneficial therapeutic properties of *Salvia officinalis* leaves in elderly patients suffering from mild to moderate Alzheimer's disease [8, 14, 23–25].

1.3.6 *Ammi visnaga*

Khella is used to treat breathing system disorders such as asthma, bronchitis, whooping cough, cardiac diseases and liver and gall bladder disorders and is believed to help discharge of kidney stones and gallstones. Khella contains coumarins and furocoumarins the most significant being khellin, visnadin and visnagin. Visnadin exhibits coronary and peripheral vasodilatory activities in isolated vascular smooth muscle of rat model. The coronary, urological and respiratory clinical and therapeutic effectiveness of khellin is well-established. Furthermore, khella has been found effective in the treatment of mild angina complaints, postoperative treatment of urinary calculus and supportive treatment of mild forms of obstructive pulmonary diseases [8, 14, 26–28].

1.3.7 *Silybum marianum*

Milk thistle is used in liver supporting functions, treatment of chronic and acute liver disease, as well as promoting detoxifying functions of the liver. The active compounds of *Silybum marianum* are flavonolignans including silidianin, silybin and silichristine, jointly recognized as silymarin. Silybin is the molecule with the greatest degree of pharmacological effects. Silymarin protects against several types of xenobiotic including alcohol. Medical studies have verified the pharmacological effects of standardized milk thistle extracts in cases of cirrhosis, toxic liver and allied liver conditions [8, 14, 29].

1.3.8 *Inula viscose*

Tayun, has been used traditionally as one of the most effective medicinal herb for the treatment of infections, inflammations, various skin diseases and wound healing. The roots have been utilized to treat cough, phlegm and sepsis. The leaves contain essential oils, flavonoids such as rhamnocitrin, glycosyl analogue of diacylglycerol, sakuranetin, methylaromadendrin, acetyl-methylaromadendrin, and sesquiterpene lactones. *Inula viscosais* well documented to have anti-ulcerogenic effects and to cause abortion in mammals. Further effects include antimicrobial and fungi effects, anti-inflammatory and anti-diabetic properties. Due to these wide range of medical effects, this plant is appreciated by pharmaceutical industries [8, 30–33].

1.3.9 *Portulaca oleracea*

Purslane, is used traditionally in the treatment of a wide range illnesses, such as include headache, stomach ache, enteritis, painful urination, mastitis. It is also used to increase milk production in nursing women and in treating postpartum bleeding.

Arial parts of this plan have therapeutic effects in healing burns, earache, ulcers, and pruritis. They are also used to treat inflammations, skin sores, insect stings, eczema, and wound healing. Purslane contains relatively high levels of a neurohormone (L-norepinephrine) that was found to exhibit vasopressor and anti-hypotensive and anti-haemorrhage effects. Various in vitro and in vivo studies indicate that aqueous extracts of this plant exhibit skeletal muscle relaxant effects [3, 4, 8].

1.3.10 *Eruca sativa*

Rucola, is considered traditionally as a general tonic and potent aphrodisiac. Additional traditional uses of *Eruca sativa* include stimulation of spermatogenesis and fertility, antibacterial effects, and promoting kidney function and digestion. *Eruca sativa* extract was found to exhibit significant antioxidant properties. Glucoerucin and flavonoids are the major antioxidants present in *Eruca sativa.* Feeding of *Eruca sativa* extract to rats induced a significant protection against HgCl induced renal toxicity. In addition, there are several scientific reports that indicate that *Eruca sativa* exhibit antimicrobial effect [8, 14, 34, 35].

1.3.11 *Cichorium intybus*

Wild chicory, finds a widespread use both in the inhibition and treatment of a wide spectrum of illnesses. It is useful in promoting liver functions (detoxifying functions) as well as encouraging the eliminative pathways both through the intestine and the kidneys. Arabic traditional healers recommend chicory as part of a combined treatment of metabolic problems, colds, and flu. The roots of chicory contain inulin and oligofructose polysaccharides. Chicory, like many plants that support liver function and immunity, has strong antioxidant effects in vitro, but the clinical significance of this has not been tested. However there have been several studies in humans on the therapeutic effects of the inulin and oligofructan polysaccharides. They have been found to undergo fermentation in the colon and to selectively stimulate of the growth of healthy bifidobacteria population that results in the decrease of colonic diseases and diabetes, as well as support for the immune system [14, 36, 37].

1.3.12 *Allium sativum, Garlic, and Onion (Allium cepa L.)*

These are one of the most used plants for their well-known health benefits. Garlic has been used for centuries for prevention and treatment of a large number of illnesses. Chinese as well as Greeks and Romans utilized onion and garlic-based preparations for the treatment and prevention of diseases. The two plants are rich sources of large number of active compounds. Onion and garlic is important ingredient of the Mediterranean diet. Various scientific reports indicate their efficacy in the treatment and prevention of a wide range of pathological conditions, such as cancer, cardiovascular diseases and obesity [3, 4, 8, 14].

1.3.13 *Urtica dioica*

Stinging nettle is highly appreciated in Greco-Arab medicine for its beneficial effects. These include anti-rheumatic effects, anti-colds and anti-cough, and promoting liver functions, anti-hypotensive and anti-inflammatory effects. Evidence-based therapeutic application of this plant includes anti-diabetes antioxidant, anti-inflammatory, increasing cell growth lymphocytes in humans, anti-prostatic hyperplasia and anti-hypertension [8, 14, 38, 39].

1.3.14 *Melissa officinalis*

Lemon balm, the therapeutic uses of this plant dates back into ancient times. Greco-Arab and Islamic physicians used the herb to treat heart disorders. *Melissa officinalis* leaves contain about 0.1 % of essential oil, consisting of a highly variable mixture of constituents. These include and polyphenolic compounds (mainly rosmarinic acid and monoterpene glycosides), monoterpenoid aldehydes, and flavonoids. Currently, *Melissa officinalis* is highly appreciated by traditional healers. The herb finds a widespread use in the treatment of skin diseases (mainly acne). Additional therapeutic effects include sedative effect on the central nervous system of mice, antimicrobial and antiviral effects, anti-hyperthyroidism and anti-depression effects. Lemon palm has also positive effects on the nervous system. In addition to anti-depression effects on patients, the plant is effective in decreasing symptoms of Alzheimer's and dementia such as memory loss [14].

1.3.15 *Pimpinella anisum*

Anise, as cumin, fennel, carrots, cilantro, and dill belong to the Apiaceae family. The seeds ("fruits") are used traditionally to treat of a wide range of illnesses, particularly for their beneficial effects in reduction of problems related to digestion. Seed-based therapies are commonly used with babies and children to heal from baby colic. Furthermore, these seeds are also recommended by traditional healers to treat symptoms associated with indigestion and nausea. In additional, their antispasmodic effect is one of most known therapeutic property of anise. The seeds commonly used to treat menstrual pain, asthma attacks, whooping cough and other spasmodic coughs. Furthermore, anise seed-based preparations are rationally used for their ability to increase the production of milk in nursing mothers. Anise-derived essential have the same therapeutical properties as the whole seeds. Women in the first term of pregnancy must should avoid taking anise [8, 14].

1.3.16 *Chamomilla recutita*

Chamomile is appreciated for the medicinal benefits of the essential oils and infusions prepared from flower heads. Due to their aromatic, flavoring and coloring properties they find a widespread use in commercial products including liniments, balms, hair products, soaps, detergents, perfumes, bakery and confectionary products, and herbal teas. Phenolic compounds, primarily the flavonoids quercetin, patuletin, apigenin and luteolin are the main active compounds of the flowers. Medicinal benefits of Chamomile-based remedies include antioxidant, antimicrobial activities and antiplatelet activity. Animal model investigations verify anti-inflammatory effects, anti-mutagenic and cholesterol-lowering effects, as well as anti-spasmotic and anxiolytic properties [8, 14].

1.3.17 *Zingiber officinale*

The rhizome (the underground stem) of the ginger is appreciated globally both as spice and for its medicinal properties to treat arthritis, colic, diarrhea, painful menstrual pains, as well as common cold and flu. The rhizomes have been utilized since ancient times as a one of the effective herbal remedies in various systems of medicine. Currently, traditional healers recommend ginger for the treatment/prevention of nausea and vomiting related with pregnancy, cancer and motion sickness. In addition, rhizome-based extracts find a wide spread use as a gastrointestinal utility for minor stomach troubles and to cure inflammations like arthritis [8, 14].

1.3.18 *Rosmarinus officinalis*

Rosemary, the areal parts of this highly aromatic plant, known for its bitter and astringent taste, are customarily utilized all over the Mediterranean region, both as cooking spice and for their medicinal properties. Rosemary contains a number of bioactive phytochemicals, such as the antioxidants carnosic acid and rosmarinic acid. Rosemary is known for its effects on muscle relaxation. Because of this property it is conventionally used to relieve digestive problems and to ease menstrual pains. A tea made from the leaves is also taken as a tonic for calming nerves and used as an antiseptic. Several studies showed that carnosic acid, found in rosemary, exhibit a strong antioxidant and antimicrobial properties [14, 40].

1.4 Administering Herbal-Based Treatment

Several preparation methods were developed in major traditional medicines are still practiced by traditional herbalists to prepare herbal-based medicines. The majority of herbal preparations are used as tea or water diluted extracts. Heating fresh or dried plant parts in a solvent result in the extraction of bioactive phytochemical. In addition, this procedure helps to reduce or even to eliminate impurities and poisons and prior to application (Table 1.1). The chemical composition and concentration of an extract is largely affected by the solvent used in the extraction. Water extracts will be rich in hydrophilic phytochemicals, oil on the other hand will absorb hydrophobic substances. Alcohol will help in extracting, both polar and un-polar compounds. Other extraction methods include the inhalation of aerosols, essential oils (Essential oils are volatile, complex, natural compounds formed by aromatic plants), tinctures (tinctures are preparations containing alcohol), capsules and tablets and vaporized plant juices or teas [1–5, 41].

1.5 Herbal Active Compounds

Plants produce metabolites as part of their normal cellular metabolic functions. These are classified as primary metabolites, present in all plants, and secondary metabolites eliciting pharmacological effects in man and animals (Table 1.2). Basic metabolism comprises all primary metabolites essential for the survival of the plant which are involved in the primary anabolic and catabolic cellular processes responsible for types of cellular activities (e.g., cell growth and differentiation). In contrast, secondary metabolites are those that found usually only in special, differentiated cells/tissues and are not necessary for the cells/tissue themselves but are important for the plant as a whole. The number of known secondary metabolites that have been discovered to date is increasing at a constant rate. Yet, it is not only plants that produce these bioactive compounds; rather, other organisms such as fungi, bacteria,

Table 1.1 Preparations methods used for oral administrations

Administration form	Preparation methods
Whole plant	Fresh juice; fresh/dried areal parts and other underground parts
Tinctures	Preparations of plant extract with varying ratios of water and alcohol
Tisanes	Hot water extracts of plants
Decoctions and teas	Made by steeping and soaking herb (leaves, flowers, stems, roots, and bark) in water for a few minutes
Vinegars	Prepared as tinctures
Syrups	Extracts of herbs made with syrup or honey
Extracts	Extracts are liquids with a lower alcohol level than tinctures
Essential oils	Essential oil extracts are usually diluted in carrier oil

Table 1.2 Secondary metabolites and their properties

Metabolites	Examples
Primary metabolites:	
Organic compounds produced in plants	Polypeptides, cellulose, amino acids, nucleic acids, mono-saccharides, and lipids
Essential for basic cell growth and differentiation	
Produced all plant tissues	
Secondary metabolites:	
Organic compounds produced in plants	Generally grouped into classes:
Do not have essential role involved in growth and differentiation	***Polyphenols*** (Widely distributed in the plant kingdom, responsible for the colors of many flowers, others are present in bark, roots and leaves that play an important role in tanning hides and skins to give leather. Yet others are simpler compounds found in most fresh fruit and vegetables
Produced in different plant families, in specific groups of plant families or in specific tissues, cell/tissue specific, produce at different developmental stages and in response to environmental stresses	***Terpenoids and steroids*** are derived biosynthetically from isopentenyldiphosphate). Over 35,000 compounds are known
	Fatty acid-derived substances and polyketides are biosynthesized from simple acyl precursors such as acetyl CoA. More than 10,000 molecules are known
	Alkaloids are derived biosynthetically from amino acids. More than 12,000 compounds are known
	Nonribosomal polypeptides are biosynthesized from amino acids
	Enzyme cofactors are coenzymes such as pyridoxal phosphate

sponges, as well as animals, are also capable of synthesizing a large number of these compounds. In general, secondary metabolites often possess interesting therapeutic properties in humans and animals, and therefore their investigation is very important. It should not be forgotten that plants synthesize these compounds as part of their own survival strategies. For example, some secondary products are pheromones used to attract insects for pollination, while others are toxins used to deter predation. Phytoalexins protect the plant against fungal or bacterial infections. Flavonoids acts as antioxidants to neutralize free radicals generated during photosynthesis. Anthocyanins may attract pollinators or seed dispersers. Alkaloids can protects against herbivore animals or insect attacks. Plants regulate their cellular metabolism in response to the present herbivores, pollinators, microorganisms, and other environmental stresses. In addition, recent evidence has pointed to additional roles for secondary metabolites in plant development. Although the term "secondary metabolites" perhaps infers a less important role for these compounds than those involved in primary metabolism, this is not the case. In fact, many essential and nonessential compounds in this group are found in plants, and even so-called "nonessential materials" can play a role in a plant's responses against abiotic and biotic stress.

In general, secondary metabolites occur as complex mixtures. The chemical composition and concentration of same plant can vary over time in response to variation in environmental conditions. Their biosynthesis can also be influenced by a variety of factors during development, in addition to stress, which makes the determination of their complete pattern essentially very difficult. Whilst secondary metabolites can occur in the tissues as active compounds, they can also be synthesized as inactive compounds that must be transformed into active products. Compounds that are biosynthesized under stress conditions are typically not detectable in unstressed tissues; when they are synthesized after the invasion of plants by various pests. The patterns of secondary metabolites will differ depending on the species. The synthesis of secondary metabolites can occur in all plant organs, including the roots, shoots, leaves, flowers, fruit, and seeds. Some metabolites are stored in specific compartments, which may be either whole organs or specialized cell types. Within these compartments the concentration of toxic secondary metabolites may be very high, so that they can exert an efficient defense against herbivores.

In order to identify or quantify a compound of interest, the metabolite must first be extracted from the plant tissues. However, the chemical properties of a material under investigation isof great importance in the development of a relevant purification scheme (Fig. 1.1). The most important issues to be taken into account include: It must be defined whether a compound or a broad range of already known should be extracted and quantified. In addition, for individual compounds, it must be determined which properties are already known, and which solvents can be used for their extraction. And finally, the purity of the compound might be important for identification and also for bioactivity assays; in this situation the metabolite must be further purified using chromatographic methods [5, 7, 40, 42].

1.6 Synergistic Actions of Foods and Phytomedicines

In contrast to synthetic drugs based upon one pure active molecule, the majority of herbal-derived medicines exert their pharmacological action via synergistic or additive pathway of a mixture active biomolecules acting at single or multiple target tissues associated with a pathophysiological pathway. In addition to desired therapeutical action these synergistic and additive effects can be advantageous by reducing negative side effects allied with the use of drugs consisting of a single pharmaceutical molecule. Additive and synergistic effects likely have their origin in the physiological and metabolic roles of secondary products in stimulating plant survival, regeneration and growth. For instance, a combination of secondary metabolites having additive or synergistic action at multiple target cell/tissue would not only guarantee efficacy in fighting wide range of pathogens and herbivores but would also reduce or even eliminate the probabilities of these pathogens developing adaptive responses or resistance [11–14].

1.7 Therapeutic Properties of Herbal-Based Active Compounds

Plants synthesize a wide range of secondary metabolites but most are derived from a few chemical motifs. These phytochemicals can have pharmacologic properties in humans and can be chemically modified to produce new medicines. Numerous herbal-derived substances have been investigated for their pharmacologic potential as new drugs. These include flavonoids, coumarins, saponins and alkaloids. Flavonoids, in particular anthrocyanins, are probably the best elucidated phytochemicals of these biomolecules due to their potent antioxidant activity. The medical benefit of numerous plant herbal-based remedies used by traditional healers, at least in part, is attributed to their effective antioxidant effects.

As above discussed, Black seed has been used for centuries in Greco-Arab and Islamic medicine for its magic healing properties as well as its disease prevention effects. Avicenna (980–1037 AC) highlighted the medical benefits of black seeds that they act as energy-booster of human body and serves to recover from fatigue and dispiritedness. Thymoquinone presents the main active molecule responsible for the biological and pharmacological properties of black seed. It was found to inhibit a wide range of pathogenic processes. For example, antioxidant, immunemodulatory, anti-cancer, hypolipidemic, and vasoconstrictive properties in cell culture systems and animal models. Additional therapeutical properties of black seeds include, inhibition of iron-dependent microsomal lipid peroxidation, cardiotoxicity induced by doxorubin in rats, drug-induced toxicity and ameliorates the anticancer effects. One of the very important pharmacological properties of thymoquinone is its high cytotoxic effects as assessed in canine osteosarcoma, colon cancer, skin cancer and prostate cancer. In contrast thymoquinone showed low cytotoxicity to normal cells. Thymoquinone also cures many multidrug-resistant types of pancreatic adenocarcinoma, human leukemia and uterine sarcoma. Furthermore, many in vitro and in vivo mechanistic studies indicate that thymoquinone induces apoptosis through affecting multiple cellular and biochemical targets. Therefore, this compound present a promising example of phytochemical that is helpful for the prevention and treatment of many types of cancer cells. This anticancer property was also supported by studies in prostate and other cancer cells. Thymoquinone was found to inhibit angiogenesis in vivo, prohibited tumor angiogenesis in a xenograft human prostate cancer model in mouse and blocked human prostate tumor growth without any side effects. Thymoquinone also exhibits anti-proliferative activity in colon and prostate tumors implanted in nude mice. Taken together, these finding show that the anticancer and cytostatic properties are due to the effect of thymoquinone on cell cycle [18, 19]. In addition, these results indicate a great potential for the development of new synthetic derivatives of thymoquinone as anticancer drugs [43–45].

Another example of potential group phytochemical is anthocyanins. This phytochemicals are one of the most abundant flavonoid compounds and one of the most widespread families of natural pigments in the plant kingdom. These pigments, present in fruits and vegetables, provide color and promote health benefits to consumers due to their antioxidant capacity. To date, more than 600 anthocyanins have been identified in the plant kingdom. The different anthocyanin absorb light at about 500 nm and are responsible for the red, blue and purple color of fruits and vegetables. All known anthocyanins conjugates are based on six anthocyanidin aglycones derived from flavylium backbone with different glycosylations and acylations. As discussed in the coming chapters of this book, many studies in cell lines, animal models and human clinical trials suggest that anthocyanins have anti-carcinogenic and anti-inflammatory activities, provides cardiovascular disease prevention, promote obesity and diabetes control benefits, and also improve visual and brain functions. Those health benefits are mainly associated with their antioxidant effects, which clearly are influenced by the molecular mechanism related to the expression and modulation of key genes.

1.8 Examples of Herbal Compounds and Pharmacological Properties

Many drugs currently listed as conventional medications are prepared from herbal-derived active compounds. The majority of these herbal-derived medicines were discovered by the study of the old traditional medical systems, namely, the Chinese, Ayurvadic, and Greco-Arab medicine. For instance, the cardiac glycoside obtained from the foxglove (*Digitalis purpurea*) are the most cited molecules of herbal-derived drugs for treatments of cardiovascular diseases. They are matchless by any synthetic or semi-synthetic medicines and have exceptional efficacy with selective cardiotonic activity. Another example is the study of cardiovascular properties of herbs that led to the discovery of reserpine over 65 years ago. Reserpine is derived from the roots of *Rauwolfia serpentine* and Vakil in 1949 reported it as a hypertensive drug. About 10 years later, reserpine was isolated said plant, its structure was elucidated and it was synthesized in labs. Later on, reserpine was found to be a potent agent in treating Parkinson disease and depression. These results stimulated further scientific research and it was observed that reserpine decreased brain nor-epinephrine, dopamine as well as serotonin. This was a break-through in research on transmitter amine defects in depression and Parkinson's disease. This was a milestone for the development of many psychoactive drugs and led to a substantial interaction between scientists and pharmacological industry.

Other examples of herbs as a source of pharmaceutical active compounds include: Vincristine is obtained from Periwinkle and used as an anti-cancer remedy. Cinchona bark is the source of malaria-fighting quinine. For centuries, herbalists prescribed *echinacea* obtained from purple coneflower to fight infection. This herb was one of the most extensively recommended medicines in the US before the discovery and synthesis of antibiotics. Now it has been confirmed that *echinacea* improves the immune system by increasing the generation of lymphocytes. Willow bark-derived salicylic acid (Aspirin) is a key anti-inflammatory, antipyretic and analgesic molecule frequently used in clinical medicine. Another example of herbal-derived medicines is opium poppy (*Papaver somniferum*)-derived morphine which is one of the early compounds used in conventional medicine systems and is a premium painkiller. The isolation of morphine from crude opium by Serturner in 1806 stirred so much research on herabal drugs that Megendie published a medical formulary in 1821, containing only pure chemical entities, hence laid paving the pathway for the use of single and pure compounds instead of medicinal plants and their extracts [43–45].

1.9 Conclusions

In parallel with the increasing in utilization of nutraceuticals and herbal-based medicines, there is also an intensive scientific research activity dealing with their safety and health beneficial effects. These include management of wide range of diseases as well as in elucidation of their action mechanisms in vitro, in vivo and in clinical studies. Plants metabolites are classified as primary metabolites, like proteins and lipids which are found in all plants, and secondary metabolites eliciting pharmacological effects in man and animals. Basic metabolism comprises all primary metabolites necessary for the survival of the cells, which are found in all plants and are responsible for the primary metabolic functions of building and sustaining plant cells. In contrast, secondary plant metabolites are those that occur usually only in special, differentiated tissues and are not necessary for the cells themselves but are important for the plant as a whole. The number of known secondary metabolites that have been discovered to date is increasing at a constant rate. Herbal-derived medicines exercise their pharmacological actions through the synergistic or additive pathway of numerous active molecules that act at single or many target tissue linked with a physiological pathway. Many drugs currently listed as conventional medications are prepared from herbal-derived active compounds. These herbal-derived drugs were discovered through the study of traditional medical systems, namely, the Chinese, Ayurvadic, and Greco-Arab medicine. Anthocyanins are one of the most abundant flavonoid compounds and one of the most widespread families of natural pigments in the plant kingdom. The following chapters will comprehensively discuss the chemistry, metabolism, in vitro and in vivo scientific literature as well as the clinical significance of anthocyanins. Regarding pharmacological properties of anthocyanins, a lot needs to be elucidated. Understand their action mechanisms in the prevention of chronic diseases, cancer, neurodegenerative diseases, and aging are still to be unveiled. As we will see in following chapters of this book, investigations regarding absorption and distribution anthocyanins are still needed. Furthermore, the effect of long-term exposure to anthocyanins is still largely uninvestigated and more in depth in vivo and clinical studies are needed in order to elucidate implications of anthocyanins in the before mentioned health-promoting effects.

References

1. Saad, B., Azaizeh, H., & Said, O. (2008). Arab herbal medicine. *Botanical Medicine in Clinical Practice, 4*, 31.
2. Saad, B., Azaizeh, H., & Said, O. (2005). Tradition and perspectives of Arab herbal medicine: A review. *Evidence-Based Complementary and Alternative Medicine, 2*(4), 475–479.
3. Saad, B., & Said, O. (2011). Herbal medicine. In B. Saad & O. Said (Eds.), *Greco-Arab and Islamic herbal medicine: Traditional system, ethics, safety, efficacy and regulatory issues* (pp. 47–71). Hoboken: Wiley.
4. Saad, B. (2014). Greco-Arab and Islamic herbal medicines, a review. *European Journal of Medicinal Plants, 4*(3), 249–258.
5. Si-Yuan, P., Shu-Feng, Z., Si-Hua, G., Zhi-Ling, Y., Shuo-Feng, Z., Min-Ke, T., et al. (2013). New perspectives on how to discover drugs from herbal medicines: CAM's outstanding contribution to modern therapeutics. *Evidence-Based Complementary and Alternative Medicine*. doi:10.1155/2013/627375.
6. Costa-Neto, E. M. (2005). Animal-based medicines: Biological prospection and the sustainable use of zootherapeutic resources. *Anais da Academia Brasileira de Ciências, 77*(1), 33–43.
7. Li, J. W.-H., & Vederas, J. C. (2009). Drug discovery and natural products: End of an era or an endless frontier? *Science, 325*(5937), 161–165.
8. Saad, B. (2015). Integrating traditional Greco-Arab and Islamic diet and herbal medicines in research and clinical practice. In I. Ramzan (Ed.), *Phytotherapies: Efficacy, safety, and regulation* (p. 142). Hoboken: Wiley.
9. Saad, B., Azaizeh, H., Abu-Hijleh, G., & Said, O. (2006). Safety of traditional Arab herbal medicine. *Evidence-Based Complementary and Alternative Medicine, 3*(4), 433–439.
10. Pormann, P. E., Savage-Smith, E., & Hehmeyer, I. (2007). *Medieval Islamic medicine*. Edinburgh: Edinburgh University Press.
11. Cragg, G. M., & Newman, D. J. (2005). Biodiversity: A continuing source of novel drug leads. *Pure and Applied Chemistry, 77*(1), 7–24.
12. Saad, B., & Said, O. (2011). The current state of knowledge of Arab herbal medicine. In B. Saad & O. Said (Eds.), *Greco-Arab and Islamic herbal medicine: Traditional system, ethics, safety, efficacy, and regulatory issues*. Hoboken: Wiley.
13. Harvey, A. L. (2008). Natural products in drug discovery. *Drug Discovery Today, 13*(19), 894–901.
14. Saad, B., & Said, O. (2011). Commonly used herbal medicines in the mediterranean. In B. Saad & O. Said (Eds.), *Greco-Arab and Islamic herbal medicine: Traditional system, ethics, safety, efficacy, and regulatory issues* (pp. 149–227). Hoboken: Wiley.
15. Salem, M. L., & Hossain, M. S. (2000). Protective effect of black seed oil from *Nigella sativa* against murine cytomegalovirus infection. *International Journal of Immunopharmacology, 22*(9), 729–740.
16. Gilani, A., Jabeen, Q., & Khan, M. A. U. (2004). A review of medicinal uses and pharmacological activities of *Nigella sativa. Pakistan Journal of Biological Sciences, 7*, 441–451.
17. Ghosheh, O. A., Houdi, A. A., & Crooks, P. A. (1999). High performance liquid chromatographic analysis of the pharmacologically active quinones and related compounds in the oil of the black seed (*Nigella sativa* L.). *Journal of Pharmaceutical and Biomedical Analysis, 19*(5), 757–762.
18. Omar, S. H. (2008). Olive: Native of Mediterranean region and health benefits. *Pharmacognosy Reviews, 2*(3), 135–142.
19. Yaseen Khan, M., Siddharth, P., Niraj, V., Amee, B., & Vimal, K. (2007). *Olea europaea*: A phyto-pharmacological review. *Pharmacognosy Reviews, 1*(1), 114–118.
20. Lansky, E. P., & Newman, R. A. (2007). *Punica granatum* (pomegranate) and its potential for prevention and treatment of inflammation and cancer. *Journal of Ethnopharmacology, 109*(2), 177–206.
21. Raju, J., Gupta, D., Rao, A. R., Yadava, P. K., & Baquer, N. Z. (2001). *Trigonella foenum graecum* (fenugreek) seed powder improves glucose homeostasis in alloxan diabetic rat tissues

by reversing the altered glycolytic, gluconeogenic and lipogenic enzymes. *Molecular and Cellular Biochemistry, 224*(1–2), 45–51.
22. Kadan, S., Saad, B., Sasson, Y., & Zaid, H. (2013). In vitro evaluations of cytotoxicity of eight antidiabetic medicinal plants and their effect on GLUT4 translocation. *Evidence-Based Complementary and Alternative Medicine, 2013*, 549345.
23. Hohmann, J., Zupkó, I., Rédei, D., Csányi, M., Falkay, G., Máthé, I., et al. (1999). Protective effects of the aerial parts of *Salvia officinalis*, *Melissa officinalis* and *Lavandula angustifolia* and their constituents against enzyme-dependent and enzyme-independent lipid peroxidation. *Planta Medica, 65*(6), 576–578.
24. Zupkó, I., Hohmann, J., Rédei, D., Falkay, G., Janicsák, G., & Máthé, I. (2001). Antioxidant activity of leaves of Salvia species in enzyme-dependent and enzyme-independent systems of lipid peroxidation and their phenolic constituents. *Planta Medica, 67*(4), 366–368.
25. Kennedy, D. O., Pace, S., Haskell, C., Okello, E. J., Milne, A., & Scholey, A. B. (2006). Effects of cholinesterase inhibiting sage (*Salvia officinalis*) on mood, anxiety and performance on a psychological stressor battery. *Neuropsychopharmacology, 31*(4), 845–852.
26. Rauwald, H. W., Brehm, O., & Odenthal, K.-P. (1994). The involvement of a Ca2+ channel blocking mode of action in the pharmacology of *Ammi visnaga* fruits. *Planta Medica, 60*(2), 101–105.
27. Carlie, G., Ntusi, N. B., Hulley, P. A., & Kidson, S. H. (2003). KUVA (khellin plus ultraviolet A) stimulates proliferation and melanogenesis in normal human melanocytes and melanoma cells in vitro. *British Journal of Dermatology, 149*(4), 707–717.
28. Hofer, A., Kerl, H., & Wolf, P. (2001). Long-term results in the treatment of vitiligo with oral khellin plus UVA. *European Journal of Dermatology, 11*(3), 225–229.
29. Ball, K. R., & Kowdley, K. V. (2005). A review of *Silybum marianum* (milk thistle) as a treatment for alcoholic liver disease. *Journal of Clinical Gastroenterology, 39*(6), 520–528.
30. Máñez, S., Recio, M. C., Gil, I., Gómez, C., Giner, R. M., Waterman, P. G., et al. (1999). A glycosyl analogue of diacylglycerol and other antiinflammatory constituents from *Inula viscosa. Journal of Natural Products, 62*(4), 601–604.
31. Ali-Shtayeh, M. S., Yaghmour, R. M., Faidi, Y. R., Salem, K., & Al-Nuri, M. A. (1998). Antimicrobial activity of 20 plants used in folkloric medicine in the Palestinian area. *Journal of Ethnopharmacology, 60*(3), 265–271.
32. Maoz, M., Kashman, Y., & Neeman, I. (1999). Isolation and identification of a new antifungal sesquiterpene lactone from *Inula viscosa. Planta Medica, 65*(3), 281–282.
33. Tripathi, Y., Tripathi, P., & Upadhyay, B. (1988). Assessment of the adrenergic beta-blocking activity of *Inula racemosa. Journal of Ethnopharmacology, 23*(1), 3–9.
34. Jirovetz, L., Smith, D., & Buchbauer, G. (2002). Aroma compound analysis of *Eruca sativa* (*Brassicaceae*) SPME headspace leaf samples using GC, GC-MS, and olfactometry. *Journal of Agricultural and Food Chemistry, 50*(16), 4643–4646.
35. Lamy, E., Schröder, J., Paulus, S., Brenk, P., Stahl, T., & Mersch-Sundermann, V. (2008). Antigenotoxic properties of *Eruca sativa* (rocket plant), erucin and erysolin in human hepatoma (HepG2) cells towards benzo (a) pyrene and their mode of action. *Food and Chemical Toxicology, 46*(7), 2415–2421.
36. Kisiel, W., & Zielińska, K. (2001). Guaianolides from *Cichorium intybus* and structure revision of *Cichorium* sesquiterpene lactones. *Phytochemistry, 57*(4), 523–527.
37. Roberfroid, M. B. (1999). Concepts in functional foods: The case of inulin and oligofructose. *The Journal of Nutrition, 129*(7), 1398S–1401S.
38. Gülçin, I., Küfrevioglu, O. I., Oktay, M., & Büyükokuroglu, M. E. (2004). Antioxidant, antimicrobial, antiulcer and analgesic activities of nettle (*Urtica dioica* L.). *Journal of Ethnopharmacology, 90*(2), 205–215.
39. Farzami, B., Ahmadvand, D., Vardasbi, S., Majin, F. J., & Khaghani, S. (2003). Induction of insulin secretion by a component of *Urtica dioica* leave extract in perifused Islets of Langerhans and its in vivo effects in normal and streptozotocin diabetic rats. *Journal of Ethnopharmacology, 89*(1), 47–53.

40. Salem, M. L. (2005). Immunomodulatory and therapeutic properties of the *Nigella sativa* L. seed. *International Immunopharmacology, 5*(13), 1749–1770.
41. Bakkali, F., Averbeck, S., Averbeck, D., & Idaomar, M. (2008). Biological effects of essential oils – A review. *Food and Chemical Toxicology, 46*(2), 446–475.
42. Gali-Muhtasib, H., Diab-Assaf, M., Boltze, C., Al-Hmaira, J., Hartig, R., Roessner, A., et al. (2004). Thymoquinone extracted from black seed triggers apoptotic cell death in human colorectal cancer cells via a p53-dependent mechanism. *International Journal of Oncology, 25*(4), 857–866.
43. Ji, H. F., Li, X. J., & Zhang, H. Y. (2009). Natural products and drug discovery. *EMBO Reports, 10*(3), 194–200.
44. Amin, A., Gali-Muhtasib, H., Ocker, M., & Schneider-Stock, R. (2009). Overview of major classes of plant-derived anticancer drugs. *International Journal of Biomedical Science, 5*(1), 1–10.
45. Briskin, D. P. (2000). Medicinal plants and phytomedicines. Linking plant biochemistry and physiology to human health. *Plant Physiology, 124*(2), 507–514.

Chapter 2
Introduction to Anthocyanins

2.1 Introduction

Medicinal plants and diet therapies are the oldest methods of prevention and treatment of all types of known diseases and built a substantial part of the main traditional medical systems in maintaining healthy body, soul, and spirit. The past three decades have witnessed significant increase in utilization as well as a big progress in studying the cellular and molecular interactions between intake of healthy foods (e.g., edible wild plants, nuts, seeds, fruits, vegetables as well as olive oil) and reduced rate of diabetes, common cancers, cardiovascular diseases, ageing and degenerative diseases. The terms functional foods and nutraceuticals refer to phytochemical substances that have long-term health beneficial effects. Although there is very thin difference between medicinal plants and functional foods, the latter also known as super-foods provide health-promoting and immunity boosting effects besides providing optimum nutrition with minimum amounts. The use of medicinal plants however leads to specific biological actions without any nutritional aspect. These pharmacological effects of medicinal plants, their extracts or their formulations are due to secondary metabolites also known as bioactive compounds or phytochemicals present in them. These secondary metabolites are specific for plant species and families and also serve a role in taxonomic identity of their respective species and families. The secondary metabolites differ from primary metabolites such as lipids, proteins, carbohydrates and nucleic acids which are commonly found in all plants and are responsible for primary metabolism and cell growth and regeneration. Various plant secondary metabolites, also known as active compounds can regulate the metabolism and physiological functions of the human body, and support immune responses. As discussed in Chaps. 7, 8, and 9, both arms of the immune system, namely the innate and the adaptive systems are indispensable for defending the body against attack by pathogens or cancer cells, and thus play a pivotal role in the maintenance of health. Hence, the intake of herbs and foods with

M. Riaz et al., *Anthocyanins and Human Health*, SpringerBriefs in Food, Health, and Nutrition, DOI 10.1007/978-3-319-26456-1_2

immunity-modulating effects are an effective way to inhibit deterioration of immunity potential and decrease the chances of allied disorders like cancer, inflammation or infection [1–3].

The term anthocyanin is coined from the Greek *anthos*, meaning flower, and *kyanos*, meaning blue in 1835 by a German scientist Ludwig Marquart as previously they were known as colored cell sap, cyanophylls, chrythophylls or cyanins. Earlier it was believed that these colors arise from the degradation of chlorophyll during autumn. Boyle in 1664 in his research paper entitled, "Experiments and considerations touching colors" for the first time noticed that color change of the "syrup of violet" from intense purple to green is due to "acid liquor". The first book ever written was, "The anthocyanin pigments of plants" by Muriel Wheldale in 1916. Willstatter and Everest (1913), Willstatter and Nolan (1915) and Shibata et al. (1919) performed first time experiments on pH dependent changes of anthocyanin colors of *Centaurea cyanus* flowers. This group reported for the first time that development of *C. cyanus* petal color is due to conjugation of anthocyanins with metal ions [4–9].

The molecular structure of anthocyanins was discovered by Richard Willstatter and his colleagues from 1912 to 1916 shortly after the rediscovery of Mendel's laws of inheritance. In this era, anthocyanins were subject of molecular genetics since Mendel's pea had distinct color due to anthocyanins [10]. The color displayed by these constituents was described by Pauling for the first time in 1939 who suggested that the resonate structure of the flavylium cation causes intensity of their color. These are the biggest group of vacuolar natural colors that are water soluble [11] and are the most important pigments followed by chlorophylls that are visible to human eye. Their color is due to their resonating structure which is also the basis of their instability. Being phytochemicals, these molecules cannot be synthesized by humans and animals and they have to rely on plants for their requirements. Anthocyanins are found in many plant tissues predominantly in the epidermal tissues, palisade, spongy mesophyll of leaves, flesh of fruits/stem and underground storage organs [12]. but most commonly accumulated in flowers and fruits [13]. Anthocyanins provide color to plant tissues (flowers, stems, leaves, roots) ranging from red, purple to blue according to the pH and their structural composition.

The anthocyanins are recently the focus of substantial basic and pharmacological research because of their sparkling color, high water solubility and valuable biological properties. They are less potent as compared to corresponding medications, but since they are consumed in substantial amounts in regular diet, they exert clearly visible long-term biological effects. The French paradox revealed that French population has less chances of coronary heart disease because of more consumption of red wine. The plethora of available flora in region and their possible combinations used in making this wine provides a diverse and rich pool of anthocyanins. These compounds play an important role in human in prevention and treatment of a wide range of diseases. Now-a-days, anthocyanins are becoming an integral part of human diet. Nutritionists and diet-counselors are increasingly recommending the use of anthocyanin-rich foods as well as pure anthocyanins to treat many diseases, due to their proven ability to simulate specific hormones and neurotransmitters, to inhibit some enzymes, and to act as antioxidant. They also act as a secondary antioxidant defense system in plant tissues exposed to different abiotic and biotic environmental stresses like fungal pathogens, UV light, cold temperature and dry weather [14].

2.2 Chemical Structure

The understanding of the chemistry of anthocyanins is of vital importance in assessment their biological and pharmacological effects. Although hundreds of anthocyanins have been identified, only a small fraction of these molecules has been studied in depth. Each anthocyanin has its specific and precise three dimensional structures which ensures optimal molecular fitting into specific cell and sub-cellular binding sites leading to biological effects. The biological properties of anthocyanins depend on their chemical structure, substitutions, conjugations and polymerization. In nature, these exist as glycosides of flavylium salts and 90 % of identified anthocyanins are based on six major anthocyanidins (aglycone). They differ from each other depending upon sugar moiety attached to the flavyliumn cation and to the acylation, hydroxylation and methoxilation pattern. Spectroscopic, molecular and functional genomic studies have exhibited remarkable similarity among all anthocyanins in sense as they all share a common skeleton. Basic skeleton structure of anthocyanins is shown in Fig. 2.1.

Chemically, anthocyanins belong to the flavonoid group that are glycosylated polyhydroxy and polymethoxy derivatives (3,5,7,39-tetrahydroxyflavylium cation) of flavilium salts, possessing a characteristic C-6 (A ring)-C-3 (C ring)-C-6 (B ring) carbon structure [14]. Structurally, these are heterosides of an aglycone unit (anthocyanidin) which is a derivative of the flavylium ion. The general formula of anthocyanins is

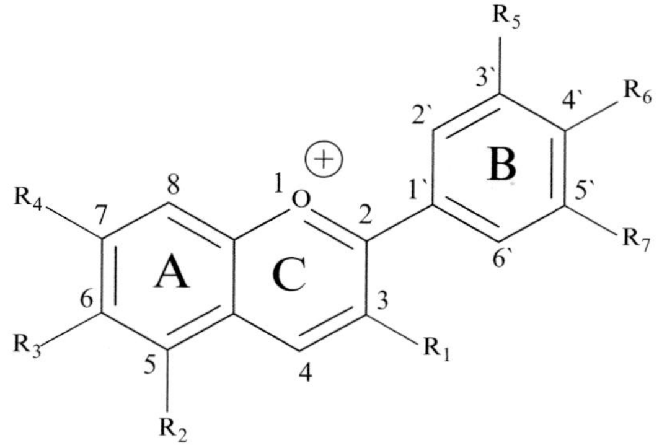

Fig. 2.1 Basic skeleton structure of anthocyanins

R_5 R_6 R_7 R_4 R_3 R_2 R_1 A B C O + 1 2 3 4 5 6 7 8 1` 2` 3` 4` 5` 6`

Anthocyanidins + sugars ⇒ Anthocyanins

The aglycone (anthocyanidins) part of the anthocyanins are quite reactive due to electron deficient flavylium cation and has lower solubility so they occur always as glycosides in nature i.e. bonded to glycosyl moiety to make it more stable than the aglycones. The most common glycone moiety are D-glucose, L-rhamnose, D-galactose, D-xylose and arabinose, these moieties are usually located at carbons 3, 5, 7, 3′, and 5′ [15, 16]. Glycosylation at C-3 position is most common as compared to other position [17]. All anthocyanins are type *O*-glycosides i.e. the sugar substituent is attached through O linkage [15]. Various factors affect their stability in food including processing and storage temperature, pH, content and identity in food matrix, oxygen, enzymes and metallic ions. Anthocyanins are different from each other by number of hydroxylated groups in the anthocyanidin, the nature and the number of bonded sugars in their structure, the aliphatic or aromatic carboxylates bonded to the sugar in the molecule, and the position of these bonds [18]. The variation of methoxylation and hydroxylation patterns in these structures produces hues from orange-red (pelargonidin) to blue-violet (delphinidin) at pH ≈ 1. Remarkably, the anthocaynins behave in same way *in vivo*: nasturtium flowers (*Tropaeolum majus* L.), radish (*Raphanus sativus* L.) and strawberry (*Fragaria ananassa* Duch.) are orange-red due to pelargonidin derivatives, while larkspur petals (*Delphinium consolida* L.), blueberries (*Vaccinium* sp.) and grapes (*Vitis* sp.) are blue-violet due to delphinidin-type anthocyanins [19]. Due to their water solubility, they are incorporated into aqueous food systems. Anthocyanins occurring in nature contain several anthocyanidins or aglycones, but only six are common in foods e.g. cyanidin, peonidin, pelargonidin, malvidin, delphinidin, and petunidin [20]. The structures of some common anthocyanins are given in Fig. 2.2.

2.2.1 Glycone Moiety and Acylating Acids

Glucose, arabinose, rhamnose, galactose, xylose or glucuronic acid are the reported monosaccharaides chains that are linked to C-3/5/7/3′/4′/5′, which are often acylated by aliphatic acids such as acetic, malonic, malic, oxalic, tartaric and succinic acids and also by cyclic acids like caffeic, sinapic, ferulic, gallic, p-hydroxybenzoic and p-coumaric acids, concurrently up to three acylating acids may be present [21]. Pendant sugars increase stabilization, and absorption of these small molecules (±500 g/mol) into human cells and their small size helps in penetrating the blood-brain barrier and cell walls. Some common monosaccharides that are linked to aglycone moiety are shown in Fig. 2.3 while structures of some acylated cyclic and aliphatic acids are depicted in Figs. 2.4 and 2.5, respectively.

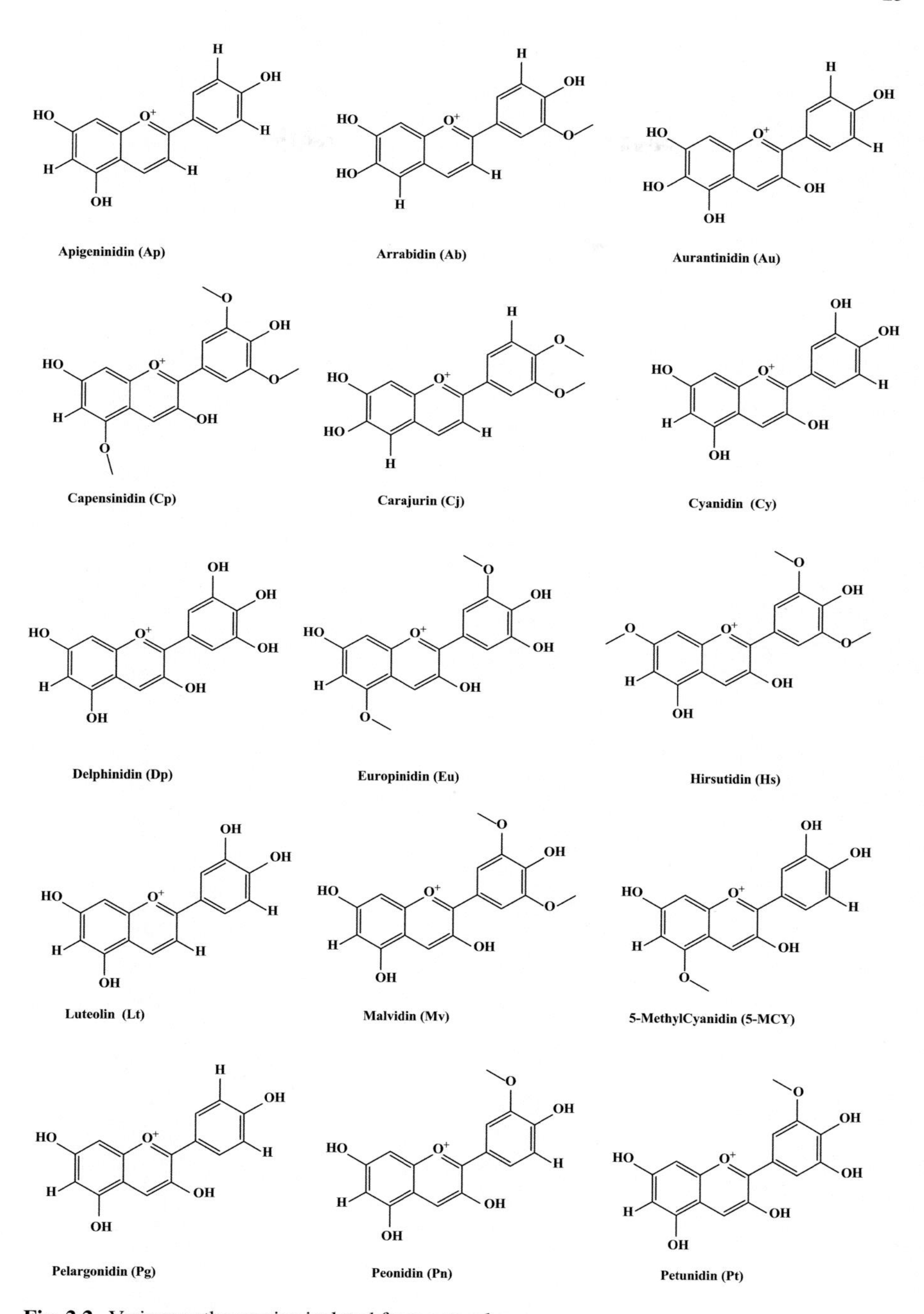

Fig. 2.2 Various anthocyanins isolated from natural sources

(continued)

Fig. 2.2 (continued)

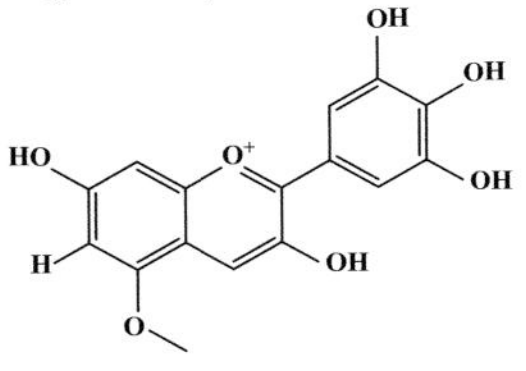

Pulchellidin (Pl)

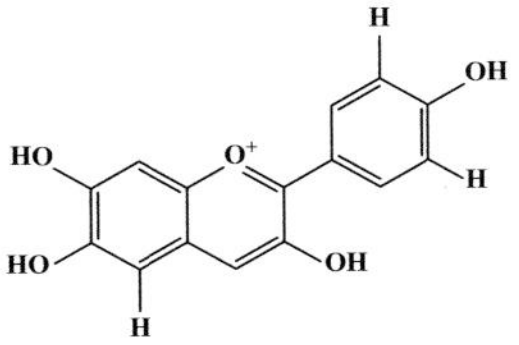

Riccionidin A (RiA)

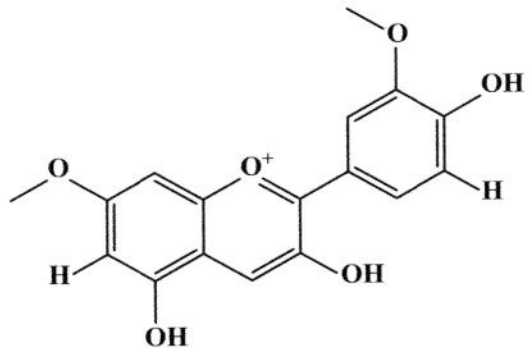

Rosinidin (Rs)

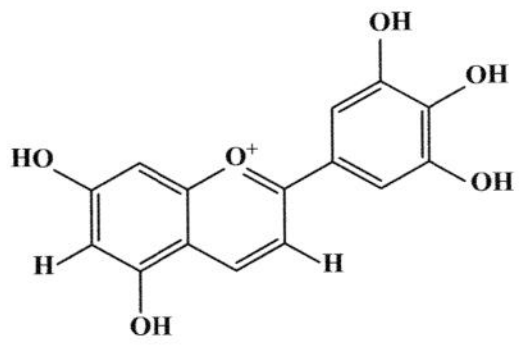

Tricetinidin (Tr)

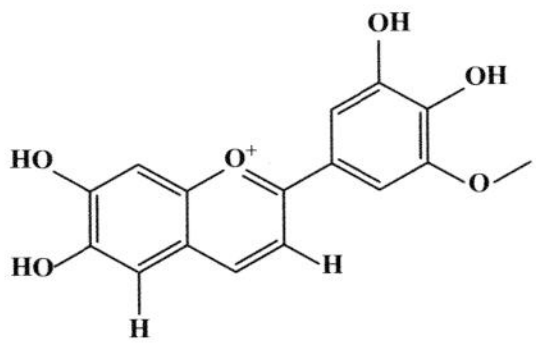

30-HydroxyAb (3'OHAb)

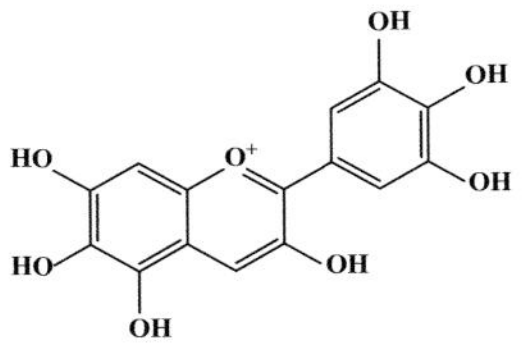

6-HydroxyCy (6OHCy)

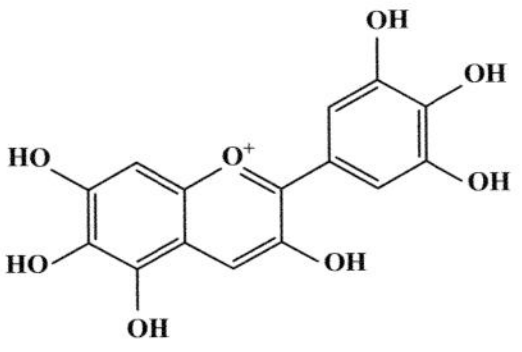

6-HydroxyDp (6OHDp)

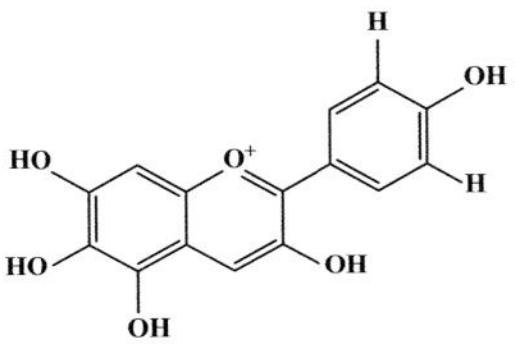

6-HydroxyPg (6OHPg)

Fig. 2.3 Monosaccharides that are linked to aglycone

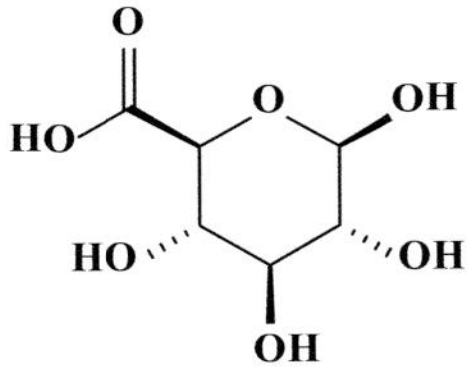

Glucuronic acid

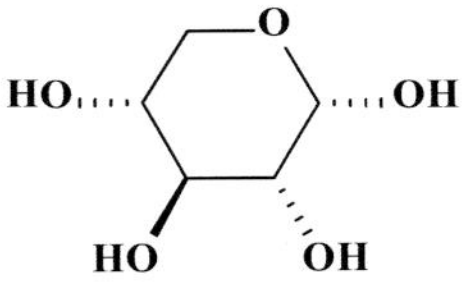

Xylose

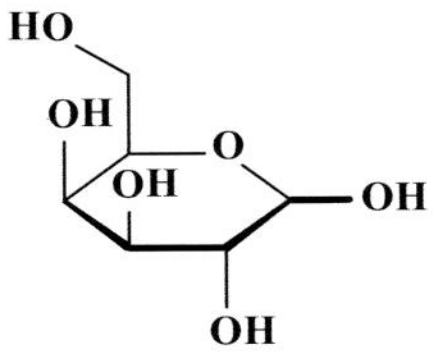

Galactose

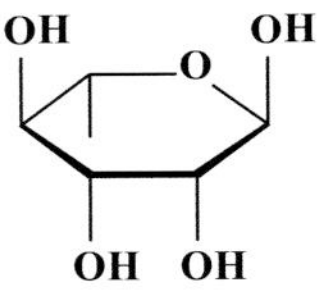

Rhamnose

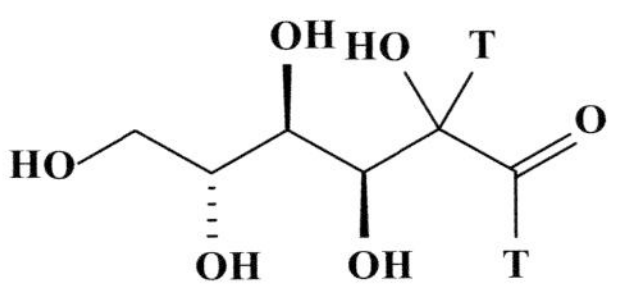

Glucose

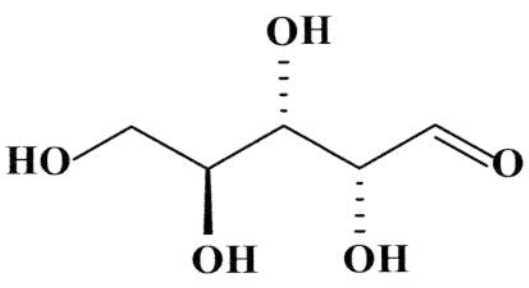

Arabinose

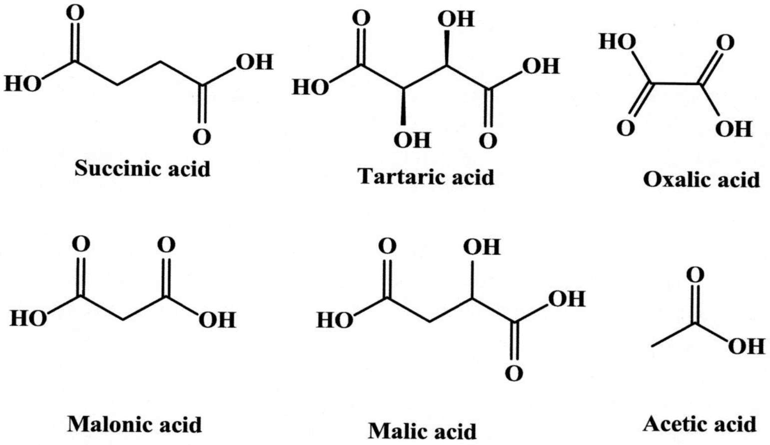

Succinic acid

Tartaric acid

Oxalic acid

Malonic acid

Malic acid

Acetic acid

Fig. 2.4 Acylated aliphatic acids

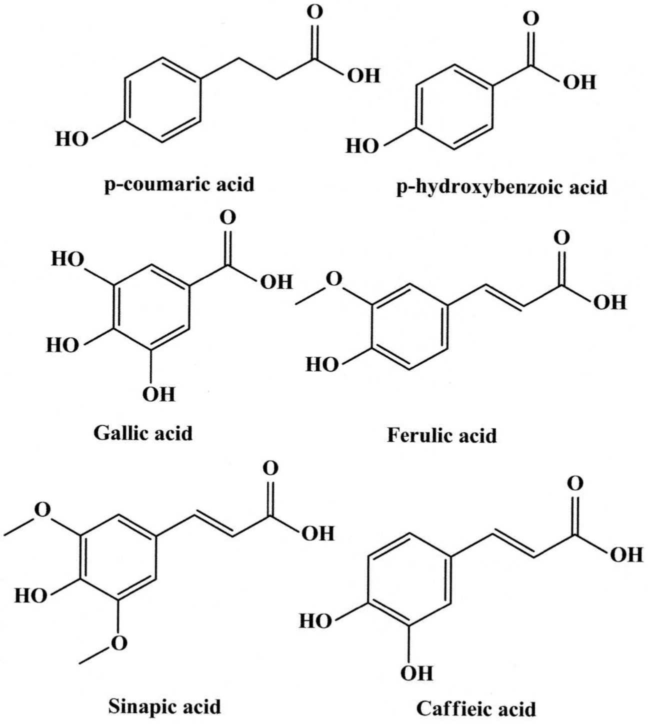

p-coumaric acid

p-hydroxybenzoic acid

Gallic acid

Ferulic acid

Sinapic acid

Caffieic acid

Fig. 2.5 Acylated cyclic acids

2.3 Pyranoanthocyanins

Pyranoanthocyanins are the products of anthocyanins and low molecular weight such as pyruvic acid, 4-vinylphenol and flavonols. They are stable derivatives of anthocyanins and considered to contribute the age-related color changes in plants, and the first pyranoanthocyanins was reported in 1996 in red wines filtrates [22]. Some pyranoanthocyanins are also detected in black carrot juice and in blood orange juices [22, 23]. Pyranoanthocyanins are stable anthocyanins derivatives that were identified in red grape wines [24]. The cyclo-addition of anthocyanins and ethylenic bond of another molecule at C-4 and C-5 results in the formation of bleach and pH change resistant, pyranoanthocyanins [25, 26]. Rosacyanin B violet was the first pyranoanthocyanidin isolated from plant petals *Rosa hybrida*.

These derivatives of anthocyanins have important qualities in terms of stability to gastro environment may results in more pharmacological active molecule and also results more stable color to change of temperature, pH or other chemical reaction like bleaching with SO_2. Figure 2.6 indicates structure of various reported pyranoanthocyanins.

For example, Vistins, a group of pyranoanthocyanins in wine with orange-red color, exhibit 11 and 14 times greater color at wine pH, than malvidin-3-*O*-glucoside [27]. Almost all reported pyranoanthocyanins have orange color except portisins which has blue color. Lu et al. (2002) isolated four pyranoanthocyanins, namely pyranodelphinin C and D and pyranocyanin C and D, possessing the pyrano[4,3,2-*de*]-1-benzopyrylium core structure from an extract of blackcurrant seeds [28]. Table 2.1 shows various reported anthocyanins.

Pyranoanthocyanins have also been observed in small quantity in extracts of red onions [21] and strawberries [38]. Four reported methyl pyranoanthocyanins isolated from black currant seeds were identified as the oxidative cyclo-addition products of the anthocyanins and extraction solvent i.e. acetone [39]. It is confirmed from studies that pyaranoanthocyanins are more stable than their parent anthocyanins, so a new strategy would be to increase the stability of anthocyanins, the importance of this group has already been discussed in previous chapters.

2.4 Conclusions

Most living matter in nature has color and a major percentage of these colors are plant-derived. Anthocyanins are incorporated in food chains because of their presence in fruits and vegetables and their corresponding beverage products like juices, jams, jellies, wines, confectionary and bakery products as food color, functional food or as a dietary supplement. Anthocyanins have aromatic rings containing polar substituent groups (–OH, $–OCH_3$) and glycosyl residues making them a polar molecule. Structurally these are benzopyran derivatives; i.e. oxygenated heterocyclic compounds while chemically these are chromophores having conjugated double

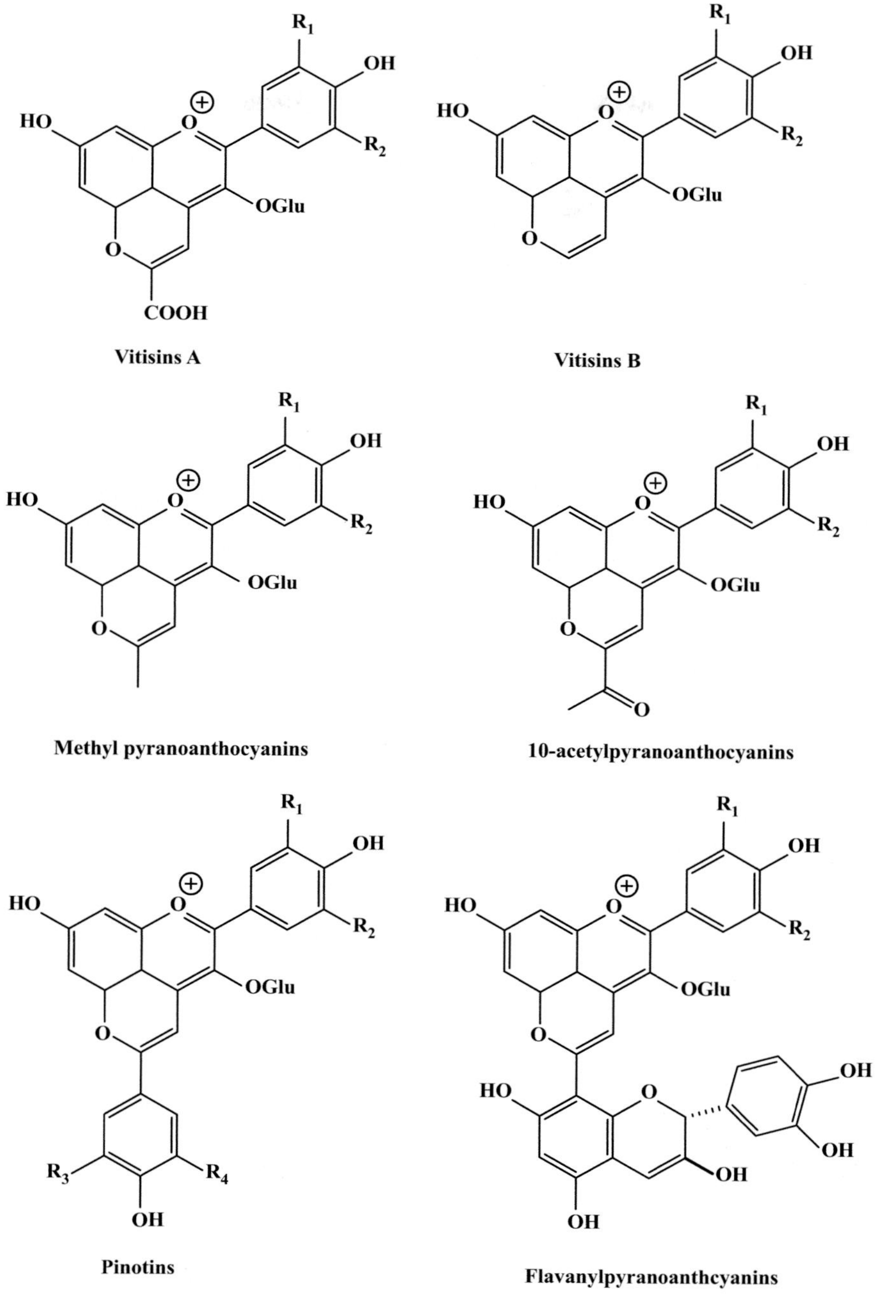

Vitisins A

Vitisins B

Methyl pyranoanthocyanins

10-acetylpyranoanthocyanins

Pinotins

Flavanylpyranoanthcyanins

Fig. 2.6 Structure of various reported pyranoanthocyanins

(continued)

Fig. 2.6 (continued)

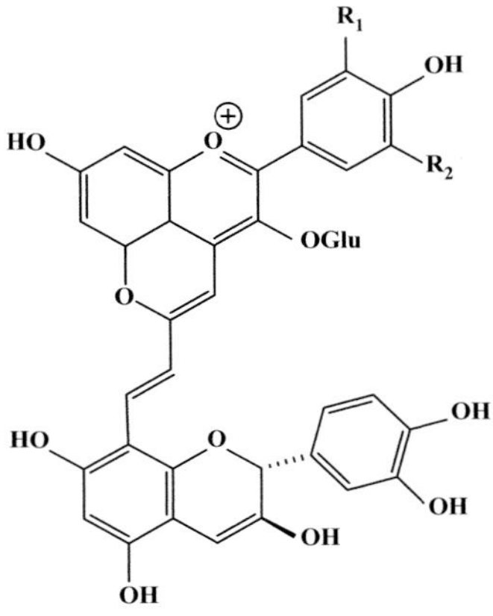

Portisins A

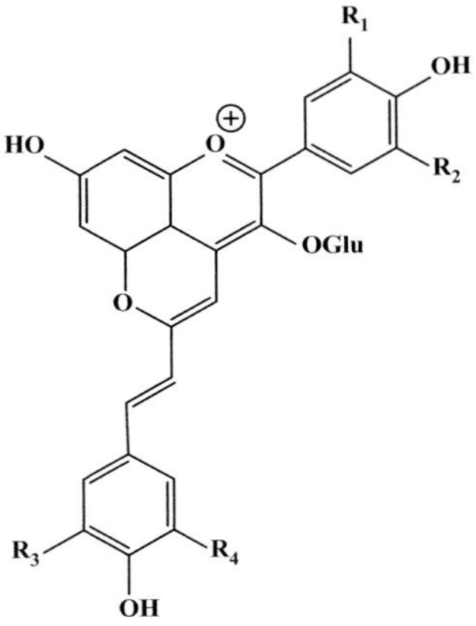

Portisins B

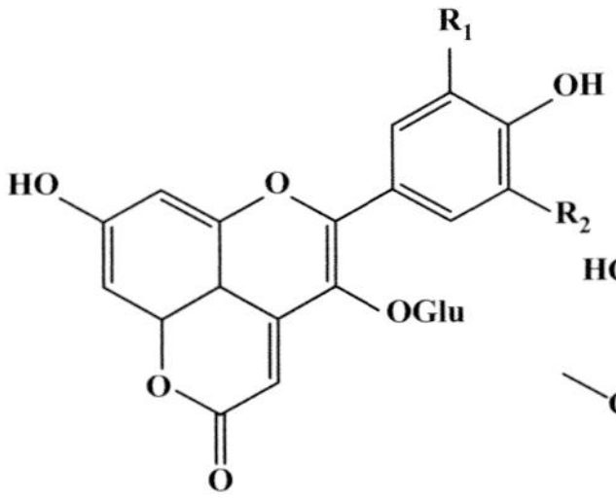

Oxovitisins

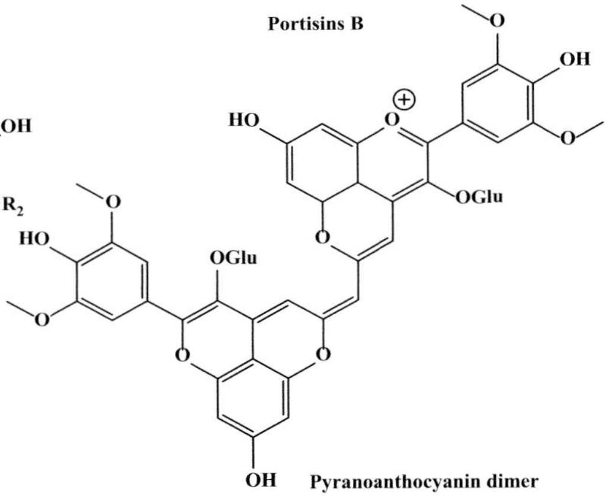

Pyranoanthocyanin dimer

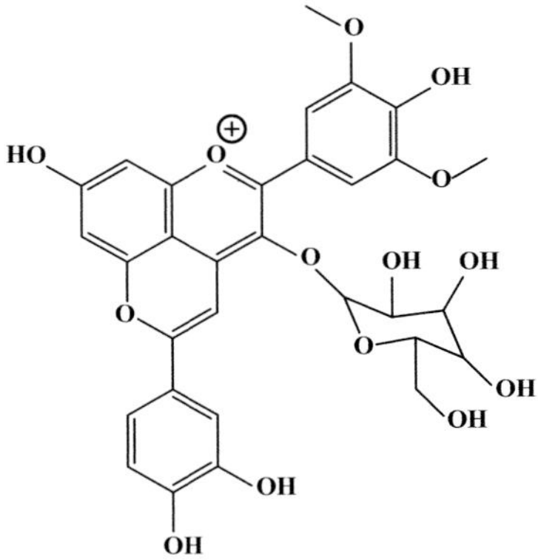

Pinotin A

Table 2.1 Various types of reported pyranoanthocyanins

Cyclo-addition	Type of pyaranoanthocyanin	References
Anthocyanin + pyruvic acid	Carboxypyranoanthocyanins (Vitisin A)	[29]
Anthocyanin + acetaldehyde	Pyranoanthocyanins (Vitisin B)	[30]
Anthocyanins + pyruvic acid + flavanols in the presence of acetaldehyde	Portisins	[31]
Anthocyanins + pyruvic acid + vinylphenol + phloroglucinol	Portisins	[32]
Anthocyanins + pyruvic acid + phenol	Portisins, oxovitisins, and pyranoanthocyanin dimmers	[33]
Anthocyanins + pyruvic acid + acetone	Methyl-pyranoanthocyanin	[34]
Anthocyanins + diacetyl	10-acetylpyranoanthocyanins	[35]
Anthocyanins + *p*-coumaric or sinapic acids	Pinotins (hydroxyphenylpyranoanthocyanins)	[36]
Pyranoanthocyanin + flavanols	Flavanylpyranoanthocyanins	[32]
Anthocyanins + 4-vinylcatechol	Pinotin A	[37]

bonds. Due to the technological advancements in instruments and chemical determinations, considerable information has been gained on the identification and characterization of anthocyanins. Despite so much progress, still hundreds of anthocyanins remain unidentified and so full potential of many plant species has not been explored yet. More work is required to isolate and characterize remaining anthocyanins and to decipher their biological activities as it will provide invaluable insights into mechanism underlying their health-promoting and immunity boosting effects.

References

1. Saad, B., & Said, O. (Eds.). (2011). *Greco-Arab and Islamic herbal medicine: Traditional system, ethics, safety, efficacy, and regulatory issues*. Hoboken: Wiley.
2. Saad, B. (2015). Integrating traditional Greco-Arab and Islamic diet and herbal medicines in research and clinical practice. In I. Ramzan (Ed.), *Phytotherapies: Efficacy, safety, and regulation* (p. 142). Hoboken: Wiley.
3. Saad, B. (2015). Greco-Arab and Islamic diet therapy: Tradition, research and practice. *Arabian Journal of Medicinal and Aromatic Plants, 1*, 1–24.
4. Boyle, R. (2004). *Experiments and considerations touching colours (1664)*. London: Henry Herringman.
5. Wheldale, M. (1916). *Anthocyanic pigments of plants*. Cambridge: Cambridge University Press.
6. Willstätter, R., & Everest, A. E. (1913). Untersuchungen über die Anthocyane. I. Über den Farbstoff der Kornblume. *Justus Liebigs Annalen der Chemie, 401*(2), 189–232.
7. Willstätter, R., & Nolan, T. J. (1915). Untersuchungen über die Anthocyane. II. Über den Farbstoff der Rose. *Justus Liebigs Annalen der Chemie, 408*(1), 1–14.

8. Shibata, K., Shibata, Y., & Kasiwagi, I. (1919). Studies on anthocyanins: Color variation in anthocyanins. *Journal of the American Chemical Society, 41*(2), 208–220.
9. Landi, M., Tattini, M., & Gould, K. S. (2015). Multiple functional roles of anthocyanins in plant-environment interactions. *Environmental and Experimental Botany, 119*, 4–17.
10. Lee, D., & Gould, K. (2002). Why leaves turn red pigments called anthocyanins probably protect leaves from light damage by direct shielding and by scavenging free radicals. *American Scientist, 90*(6), 524–528.
11. Ghosh, D., & Konishi, T. (2007). Anthocyanins and anthocyanin-rich extracts: Role in diabetes and eye function. *Asia Pacific Journal of Clinical Nutrition, 16*(2), 200.
12. Oren-Shamir, M. (2009). Does anthocyanin degradation play a significant role in determining pigment concentration in plants? *Plant Science, 177*(4), 310–316.
13. Delgado-Vargas, F., & Paredes-López, O. (2003). Anthocyanins and betalains. *Natural Colorants for Food and Nutraceutical* 167–219.
14. Chalker Scott, L. (1999). Environmental significance of anthocyanins in plant stress responses. *Photochemistry and Photobiology, 70*(1), 1–9.
15. Da Costa, C. T., Horton, D., & Margolis, S. A. (2000). Analysis of anthocyanins in foods by liquid chromatography, liquid chromatography-mass spectrometry and capillary electrophoresis. *Journal of Chromatography A, 881*(1), 403–410.
16. Strack, D., Steglich, W., & Wray, V. (1993). Betalains. *Methods in Plant Biochemistry, 8*, 421–450.
17. Bkowska-Barczak, A. (2005). Acylated anthocyanins as stable, natural food colorants – A review. *Polish Journal of Food and Nutrition Sciences, 14*, 107–116.
18. Valls, J., Millán, S., Martí, M. P., Borràs, E., & Arola, L. (2009). Advanced separation methods of food anthocyanins, isoflavones and flavanols. *Journal of Chromatography A, 1216*(43), 7143–7172.
19. Stintzing, F. C., & Carle, R. (2004). Functional properties of anthocyanins and betalains in plants, food, and in human nutrition. *Trends in Food Science and Technology, 15*(1), 19–38.
20. Kong, J. M., Chia, L. S., Goh, N. K., Chia, T. F., & Brouillard, R. (2003). Analysis and biological activities of anthocyanins. *Phytochemistry, 64*(5), 923–933.
21. Andersen, Ø. M., Fossen, T., Torskangerpoll, K., Fossen, A., & Hauge, U. (2004). Anthocyanin from strawberry (*Fragaria ananassa*) with the novel aglycone, 5-carboxypyranopelargonidin. *Phytochemistry, 65*(4), 405–410.
22. Rentzsch, M., Schwarz, M., & Winterhalter, P. (2007). Pyranoanthocyanins – An overview on structures, occurrence, and pathways of formation. *Trends in Food Science and Technology, 18*(10), 526–534.
23. Castañeda-Ovando, A., Pacheco-Hernández, M., Páez-Hernández, M. E., Rodríguez, J. A., & Galán-Vidal, C. (2009). Chemical studies of anthocyanins: A review. *Food Chemistry, 113*(4), 859–871.
24. Pozo-Bayón, M. A., Monagas, M., Polo, M. C., & Gómez-Cordovés, C. (2004). Occurrence of pyranoanthocyanins in sparkling wines manufactured with red grape varieties. *Journal of Agricultural and Food Chemistry, 52*(5), 1300–1306.
25. Monagas, M., Bartolomé, B., & Gómez-Cordovés, C. (2005). Evolution of polyphenols in red wines from *Vitis vinifera* L. during aging in the bottle. *European Food Research and Technology, 220*(3–4), 331–340.
26. Bakker, J., & Timberlake, C. F. (1997). Isolation, identification, and characterization of new color-stable anthocyanins occurring in some red wines. *Journal of Agricultural and Food Chemistry, 45*(1), 35–43.
27. Romero, C., & Bakker, J. (1999). Interactions between grape anthocyanins and pyruvic acid, with effect of pH and acid concentration on anthocyanin composition and color in model solutions. *Journal of Agricultural and Food Chemistry, 47*(8), 3130–3139.
28. Lu, Y., Foo, L. Y., & Sun, Y. (2002). New pyranoanthocyanins from blackcurrant seeds. *Tetrahedron Letters, 43*(41), 7341–7344.
29. Fulcrand, H., Montserrat, D., Erika, S., & Véronique, C. (2006). Phenolic reactions during winemaking and aging. *American Journal of Enology and Viticulture, 57*(3), 289–297.

30. Bakker, J., & Timberlake, C. F. (1985). The distribution of anthocyanins in grape skin extracts of port wine cultivars as determined by high performance liquid chromatography. *Journal of the Science of Food and Agriculture, 36*(12), 1315–1324.
31. Mateus, N., Oliveira, J., Haettich-Motta, M., & de Freitas, V. (2004). New family of bluish pyranoanthocyanins. *BioMed Research International, 2004*(5), 299–305.
32. de Freitas, V., & Mateus, N. (2011). Formation of pyranoanthocyanins in red wines: A new and diverse class of anthocyanin derivatives. *Analytical and Bioanalytical Chemistry, 401*(5), 1463–1473.
33. Schwarz, M., Wabnitz, T. C., & Winterhalter, P. (2003). Pathway leading to the formation of anthocyanin-vinylphenol adducts and related pigments in red wines. *Journal of Agricultural and Food Chemistry, 51*(12), 3682–3687.
34. Mateus, N., Pascual-Teresa, S., Rivas-Gonzalo, J. C., Santos-Buelga, C., & Victor de Freitas, V. D. (2002). Structural diversity of anthocyanin-derived pigments in port wines. *Food Chemistry, 76*(3), 335–342.
35. Gómez-Alonso, S., Collins, V. J., Vauzour, D., Rodríguez-Mateos, A., Corona, G., & Spencer, J. P. E. (2012). Inhibition of colon adenocarcinoma cell proliferation by flavonols is linked to a G2/M cell cycle block and reduction in cyclin D1 expression. *Food Chemistry, 130*(3), 493–500.
36. He, F., Liang, N. N., Mu, L., Pan, Q. H., Wang, J., & Reeves, M. J. (2012). Anthocyanins and their variation in red wines I. Monomeric anthocyanins and their color expression. *Molecules, 17*(2), 1571–1601.
37. Schwarz, M., Jerz, G., & Winterhalter, P. (2015). Isolation and structure of Pinotin A, a new anthocyanin derivative from Pinotage wine. *VITIS-Journal of Grapevine Research, 42*(2), 105.
38. Fossen, T., & Andersen, Ø. M. (2003). Anthocyanins from red onion, (*Allium cepa*) with novel aglycone. *Phytochemistry, 62*(8), 1217–1220.
39. Lu, Y., & Foo, L. Y. (2001). Unusual anthocyanin reaction with acetone leading to pyranoanthocyanin formation. *Tetrahedron Letters, 42*(7), 1371–1373.

Chapter 3
Occurrence of Anthocyanins in Plants

3.1 Introduction

Currently, there are plenty of scientific studies dealing with biological and pharmacological properties of anthocyanins as well as with their structure, composition, and abundance in the plant kingdom. Anthocyanins, are considered the most important group of flavonoids in plants having more than 600 compounds identified in nature. These water-soluble compounds are widely distributed in plant tissues and provide color to leaves, stems, roots, flowers and fruits ranging from red, purple to blue according to the environmental pH and their chemical structure. Due to their chemical structure anthocyanins are soluble in water solutions and frequently found in the vacuoles of epidermal cells, and in some species, they are bound to membrane of the main cell-vacuole, called anthocyanoplasts. The relative abundance, composition, and chemical structure of anthocyanins may vary according to the species or from fruit to fruit of the same species depending on external and internal dynamics. Genetic factors and agronomic practices, intensity and type of light, temperature, processing season, cultivation and horticultural practices, stage of maturity at harvest and post-harvest storage conditions. Thus identifying absolute anthocyanins identity and content of specific plant-derived diet is very difficult and therefore range of values is used usually. Figure 3.1 shows anthocyanins-based colors of various storage conditions influence the level of anthocyanins. Although, anthocyanins are found in almost all members of the plant kingdom, plants belonging to 27 families are rich in anthocyanins [1].

Rich edible sources of anthocyanins are colored fruits such as berries, cherries, peaches, grapes, pomegranates, and plums as well as many dark-colored vegetables such as black currant, red onion, red radish, black bean, eggplant, purple corn, red cabbage, and purple sweet potato [2]. All these fruits and vegetables are regularly consumed either in diets or in juices, soft drinks, alcoholic beverages, and similar other products [3]. Figure 3.2 summarizes the main known biological functions of anthocyanins.

M. Riaz et al., *Anthocyanins and Human Health*, SpringerBriefs in Food, Health, and Nutrition, DOI 10.1007/978-3-319-26456-1_3

Fig. 3.1 Examples of anthocyanins colored fruits, vegetable, leaves and flowers

Almost all highly pigmented fruits and their products as well as byproducts are potential sources of anthocyanins colorants, such as cranberry press cake, red-grapes extracts and their by-products [4], blueberries [5], elderberries, *Hibiscus calyces*, black choke berries, purple corn, and black current berries etc. However, the anthocyanins composition of all of these comprises chiefly mono and di-glucosides which stabilize against pH changes and hydration in a limited manner [6].

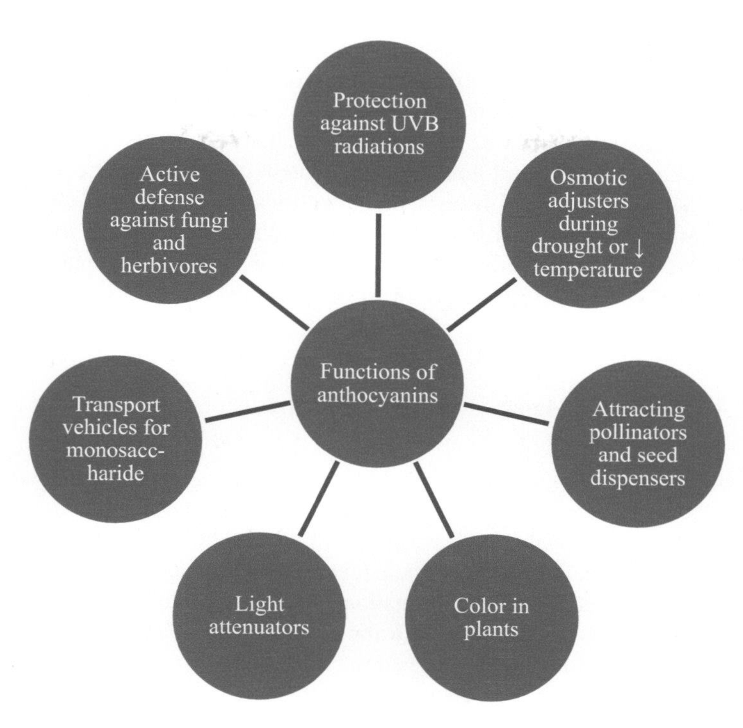

Fig. 3.2 Functions of anthocyanins in plants

As discussed in Chap. 1, anthocyanins impart shiny blue, orange, violet, pink and red colors (Fig. 3.1) to the flowers, vegetables, beverages and fruits of some plants thus these less stable and more water soluble metabolites than carotenoids, can be extracted from tulips, roses and orchids [7]. In addition, *Heliophila coronopifolia* from cape areas has blue flowers [8] and 'Dalicha' (*Tali camellia*) from Yunnan has red flowers because of anthocyanins [9] similarly *Rosa hybrida*, (*Rosaceae*), has colored appearance because of anthocyanins and carotenoids [10, 11]. Saito et al. (2011) reported 44 floral anthocyanidin glycosides of pelargonidin, cyanidin and peonidin from the Japanese morning glory (*Ipomoea nil* or *Pharbitis nil*) [8].

3.2 Concentration of Anthocyanins in Fruits, Vegetable and Nuts

As mentioned above, contents and composition of anthocyanins in plant tissue are affected by genetics (variety and cultivar), geographic origin, growing seasons, agronomic practices utilized during farming, collection time, climate (temperature, light, altitude) and geological conditions of area from where that food commodity or plant part (vegetable, fruit or flower) is collected, processing method and the method used for determination. Knowledge on individual and population level anthocyanins intake relies on precise information about their level in foods. Therefore source and content of anthocyanins in food is of vital importance for researcher, nutrition-counselors, food processors, manufacturers and consumers. Several tables and databases of anthocyanins content of foods have been compiled to facilitate anthocyanins classification and quantification of intake (Tables 3.1 and 3.2). Due to the important role in food industry and human nutrition, the common food of the United States, fruit, vegetable and nuts are assayed for total anthocyanins content per 100 g of fresh weight or form consumed (Table 3.1).

3.2.1 Variation in Anthocyanins Content

Anthocyanins are present at cellular level in vacuoles, at organ level in flowers epidermal cells and occasionally mesophylls, however in certain leaves they are restricted to mesophylls only e.g. leaves of *Secale cereal* [40]. Different food matrices require different methods for complete release of anthocyanins without degradation and therefore accurate determination of anthocyanins contents requires optimum and validated procedures both for isolation and quantification. Even for same food matrix, various factor like varietal differences, time of harvest, farming practices and raw versus processed food influence the anthocyanin contents. Sometimes many of above mentioned variables affect anthocyanin contents and

Table 3.1 Anthocyanins content of commonly consumed fruits and vegetables in United States [12]

Fruit	Total anthocyanins (mg)	Fruit	Total anthocyanins (mg)
Apple (Fuji, Gala, Red delicious)	1.8, 3.2 and 17	Marion blackberry	433
Blackberry	353	Blueberry cultivated	529
Gooseberry (average of various cultivars)	6.43	Elderberry	1993
Chokeberry	2147	Cranberry	133
Peach	4.7	Concord grape	192
Nectarine	9.2	Red grape	42.7
Currant black,	533	Currant red	14.3
Cherry (sweet)	177	Blueberry wild	705
Strawberry (average)	52.1	Raspberry black, red	845, 116
Black plum	82	Plum	12.5
Blackbeans	23.1	Red cabbage	113
Eggplant	35.1	Red leaf lettuce	1.5
Red onion	38.8	Small red bean	6.2
Red radish	116	Pistachio	2.1

mg per 100 g fresh weight

Table 3.2 Plants as sources of anthocyanins

Anthocyanins sources	Reference	Anthocyanins sources	Reference
Vegetables	[13]	Elderberries	[14]
Potato	[15]	*Hibiscus calyces*	[16]
Tomato	[17]	Black currents	[18]
Spinach	[19]	Purple corn	[20]
Paprika	[21]	Berries	[22]
Purple corn	[23]	Strawberry	[24]
Red radish	[25]	Mulberry	[26]
Red and green cabbages	[27, 28]	Tamarillo fruit (*Solanum betaceum* Cav.)	[29]
Fruits		Garcinia indica Choisy	[30]
Cherries	[31]	Litchi	[32]
Extracts of red grapes and its by-products	[33]	Apple skin	[34]
Cranberry press cake	[35, 36]	Pomegranate (*Punica granatum* L.)	[37]
Blueberries	[5]	Muscadine grapes (*Vitis rotundifolia*)	[38]
Black chokeberries	[39]		

while it may be possible to make generalization about one factor, in reality numerous factors exert their influence during synthesis and accumulation of anthocyanins. Further their post-harvest stability is affected by processing, cooking, heat-treatment, exposure to light and oxygen and storage which not only affects content but also form of anthocyanins. The concentration of anthocyanins is variable, it was reported to be 0.1–1 % per dry weight for most of the fruits and vegetable, it may be present as single main type e.g. Dianthus, apple, cherry, two main type e.g. cranberry and many/mixture of anthocyanins e.g. *Rosa*, *Tulipa*, grapes [40]. Table 3.3 shows various rich sources of anthocyanins.

Andersen and Jordheim (2006) reported that there are more than 500 different anthocyanins [47] and 23 anthocyanidins [47, 48]; of which only six are the most common in vascular plants, pelargonidin, peonidin, cyanidin, malvidin, petunidin and dephinidin [49]. The glycoside derivatives of the cyanidin, dephinidin and pelargonidin, i.e. three non-methylated anthocyanidins, are mostly found in nature, with 50 %,69 % and 80 % in flowers, fruits and leaves respectively [50].

The distribution of the six common anthocyanidins in fruits and vegetables in the abundance order as: cyanidin (50 %), pelargonidin (12 %), peonidin (12 %), delphinidin (12 %), petunidin (7 %), and malvinidin (7 %), while publishing another report he included more anthocyanins sources and the abundance order was likely to be cyanidin (30 %), delphinidin (22 %), pelargonidin (18 %), peonidin (7.5 %), malvidin (7.5 %), and peunidin (5 %). In both reports, the three non-methylated anthocyanidins (cyanidin, delphinidin, and pelargonidin) were exposed to be more prevalent than the three methylated anthocyanins (peonidin, malvidin, and petunidin). So it was confirmed that cyanidin-3-glucoside is the most common anthocya-

Table 3.3 Anthocyanins content in vegetables and fruits

Sources	Cultivar	Total anthocyanin content (mg/100 g)	Reference
Blackberries, raw	*Rubus fruticosus*	100.61	[41]
Blueberries, raw	*Vaccinium myrtillus*	163.30	[41]
Cranberry, raw	*Vaccinium oxycoccos*	103.70	[41]
Gooseberry, raw	*Ribesuva-crispa*	9.51	[41]
Raspberry, raw	*Rubus idaeus*	48.63	[41]
Strawberry, raw	*Fragaria xananassa*	27.01	[41]
Grape, raw	*Vitis vinifera*	48.04	[41]
Grape, dried	*Sunbelt*	107.60	[42]
Grape, raw, skin *Black*	*Olympia*	97.50	[43]
Pomegranate, raw	*Mollar de Elche*	56.09	[43]
Purple tomato, dried	*V118*	72.31	[44]
Purple carrot, dried	*Rain*	44.27	[45]
Purple carrot, dried	*Haze*	57.22	[45]
Red cabbage, dried	*Gario*	198.61	[45]
Purple cauliflower, dried	*Graffitti*	201.11	[45]
Purple potato, dried	*Majesty*	97.71	[45]
Purple potato, dried	*Mackintosh*	48.74	[45]
Red potato, dried	*Y38*	85.23	[45]
Red potato, dried	*Thumb*	43.09	[45]
Red onion, dried	*Pier-c*	7.77	[45]
Red onion, dried	*Pearl*	18.95	[45]
Red onion, raw	*Morada de Amposta*	23.30	[46]
Eggplant, dried	*Black Beauty*	29.55	[45]

nins in nature and more than 90 % of anthocyanins contain glucose as a glycosylating sugar [7, 40].

3.3 Typical New Anthocyanins Found in the Past Years

Four anthocyanins with furanose sugar (apiose) were obtained from leaves of *Synadenium grantii* (*Euphorbiaceae*) [51], which is very unusual in anthocyanin family. In 2003 and 2004, Honda and Tatsuzawa et al. (2004) discovered two novel C-glycosyl anthocyanins, which were isolated from the flowers of *Tricyrtis formosana* cultivar Fujimusume [52]. Pyranoanthocyanins are a newer member with a pyran ring in the anthocyanins family, which were reported in 1996 as the pigments detected in red wines filtrates [53]. Furthermore, there are some unique anthocyanin–flavanol condensed pigments (polymeric anthocyanins) reported in existing studies [54, 55]. Table 3.4 shows major and minor anthocyanins in various food while Table 3.5 shows the occurrence of anthocyanins in common Mediterranean diet.

Table 3.4 Occurrence pattern of anthocyanins in common vegetables and fruits [56]

Name	Major anthocyanins	Minor anthocyanins
Strawberry	Pelargonidin-3-glucoside	Cyanidin-3-glucoside, Pelargonidin-3-rutinoside
Blackberry	Cyanidin-3-glucoside	Cyanidin-3-rutinoside, malvidin-3-glucoside
Raspberry	Cyanidin-3-glucoside	Pelargonidin-3-glucosides, Pelargonindin-3-rutinoside
Sweet cherries	Cyanidin-3-rutinoside	Cyanidin-3-glucoside, Peonidin-3-rutinoside
Blackcurrant	Cyanindin-3-rutinoside	Cyanindin-3-glucoside, Delphinidin-3-glucoside
Bilberry	Delphinidin-3-galactoside	Peonindin-3-glucoside, Peonindin-3-galactoside
Red onions	Cyanidin 3-glucoside	Delphinidin-3-glucoside, Petunidinglucoside
Blood orange	Cyanidin 3-glucoside	Delphinidin 3-glucoside, cyanidin 3,5-diglucoside, cyanidin 3-sophoroside, delphinidin 3-(6″-malonylglucoside), peonidin 3-(6″-malonylglucoside), cyanidin 3-(6″-dioxalylglucoside)

Table 3.5 The occurrence of anthocyanins in common Mediterranean diet

Plant	Anthocyanins	References
Blackberries (*Rubus ulmifolius Schott*)	Cyanidin-3-glucoside (major) and cyanidin-3-rutinoside (minor)	[57]
Eggplant (*Solanumm elongena L.*)	Nasunin	[58]
Fig (*Ficus carica L.*)	Cyanidin-3-*O*-rhamnoglucoside (cyanidin-3-*O*-rutinoside) (major), cyanidin 3-glucoside, cyanidin 3,5-diglucoside and pelargonidin 3-rutinoside	[59, 60]
Grape (*Vitis vinifera L.*) and Wine	3-*O*-monoglucoside of pelargodin-3-*O*-glucoside, cyanidin-3-*O*-glucoside, delphinidin-3-*O*-glucoside, peonidin-3-*O*-glucoside, petunidin-3-*O*-glucosideand malvidin-3-*O*-glucoside	[61]
Lentil (*Lens culinaris Medik.*)	Delphinidin-3-*O*-(2-*O*-β-D-glucopyranosyl-α-L-arabino-pyranoside	[62]
Myrtle (*Myrtus communis L.*)	Anthocyanin arabinosides	[63]
Olive (*Olea europea L.*)	Cyanidin 3-*O*-glucoside and cyanidin 3-*O*-rutinoside	[64]
Sweet Cherry (*Prunu savium L.*)	Cyanidin 3-rutinoside, cyanidin 3-glucoside, peonidin 3-rutinoside, cyanidin 3-sophoroside, pelargonidin 3-glucoside, pelargonidin 3-rutinoside, 3-glucoside	[65, 66]
Plum (*Prunus domestica L.*)	Cyanidin-3-glucoside, cyanidin-3-rutinoside, peonidin-3-glucoside, and peonidin-3-rutinoside	[67, 68]

(continued)

Table 3.5 (continued)

Plant	Anthocyanins	References
Onion (*Allium cepa L.*)	3-(3-Glucosyl-6-malonyl-glucoside), 3-(6-malonyl-glucoside), 3-(3-glucosylglucoside) and cyanidin-3-glucoside, respectively 3-(3,6-dimalonylglucoside), 3-(3-malonylglucoside), and 3,5-diglucoside of cyanidin, 3-glucoside, 3,5-diglucoside and peonidin-3-malonyl-glucoside	[69, 70]
Red Radish (*Raphanus sativus L.*)	Cyanidin and pelargonidin derivatives pelargonidin-3-sophoroside-5-glucoside derivative compounds were found	[71]
Red chicory (*Cichorium intybus L. var. Silvestre Bisch.*)	Cyanidin-3-*O*-(6-malonyl-β-glucopyranoside) delphinidin derivative has been recently identified	[72, 73]
Pomegranate (*Punica granatum L.*)	3-Glucosides and 3,5-diglucoside such as delphinidin-3-glucoside, cyanidin-3,5-diglucoside, delphinidin-3-glucoside and pelargonidin-3-glucoside	[74, 75]

3.4 Conclusions

Various fruits, vegetables, nuts, herbs, spices and horticultural crops are good source of anthocyanins. Diverse factors like packaging, freezing, lipid composition and other bioactive constituents present in fruits and vegetables affect anthocyanin contents. Release of organic acids that can lead to isomerization during slicing and juicing of fruits, surface area, porosity and enzymatic oxidation during slicing, peeling, pulping or juicing all affect anthocyanins contents in fruits. Therefore marketing minimally processed fruits and vegetables is continuously increasing due to consumer demand for high quality and nutritive products. The development of novel and more ample sources of anthocyanins in horticultural crops, vegetables and fruits and improvement of their shelf-life and consumer acceptance can improve human health significantly. The ongoing research on horticulture and crops is expected to bring more verities on grocery shelves and in hands of consumers having more anthocyanins contents than traditional varieties.

References

1. Ghosh, D., & Konishi, T. (2007). Anthocyanins and anthocyanin-rich extracts: Role in diabetes and eye function. *Asia Pacific Journal of Clinical Nutrition, 16*(2), 200.
2. Wu, X., Pittman, H. E., & Prior, R. L. (2006). Fate of anthocyanins and antioxidant capacity in contents of the gastrointestinal tract of weanling pigs following black raspberry consumption. *Journal of Agricultural and Food Chemistry, 54*(2), 583–589.
3. Valls, J., Millán, S., Martí, M. P., Borràs, E., & Arola, L. (2009). Advanced separation methods of food anthocyanins, isoflavones and flavanols. *Journal of Chromatography A, 1216*(43), 7143–7172.

4. Camire, A., & Clydesdale, F. (1979). High pressure liquid chromatography of cranberry anthocyanin. *Journal of Food Science, 44*(3), 926–927.
5. Francis, F. (1985). Blueberries as a colorant ingredient in food products. *Journal of Food Science, 50*(3), 754–756.
6. Giusti, M. M., & Wrolstad, R. E. (2003). Acylated anthocyanins from edible sources and their applications in food systems. *Biochemical Engineering Journal, 14*(3), 217–225.
7. Castañeda-Ovando, A., Pacheco-Hernández, M., Páez-Hernández, M. E., Rodríguez, J. A., & Galán-Vidal, C. A. (2009). Chemical studies of anthocyanins: A review. *Food Chemistry, 113*(4), 859–871.
8. Saito, N., Tatsuzawa, F., Toki, K., Shinoda, K., Shigihara, A., & Honda, T. (2011). The blue anthocyanin pigments from the blue flowers of *Heliophila coronopifolia* L. (Brassicaceae). *Phytochemistry, 72*(17), 2219–2229.
9. Li, J. B., Hashimoto, F., Shimizu, K., & Sakata, Y. (2008). Anthocyanins from red flowers of *Camellia cultivar* 'Dalicha'. *Phytochemistry, 69*(18), 3166–3171.
10. Lee, J. H., & Choung, M.-G. (2011). Identification and characterisation of anthocyanins in the antioxidant activity-containing fraction of *Liriope platyphylla* fruits. *Food Chemistry, 127*(4), 1686–1693.
11. Lee, J. H., Lee, H.-J., & Choung, M.-G. (2011). Anthocyanin compositions and biological activities from the red petals of Korean edible rose (*Rosa hybrida* cv. Noblered). *Food Chemistry, 129*(2), 272–278.
12. Wu, X., Beecher, G. R., Holden, J. M., Haytowitz, D. B., Gebhardt, S. E., & Prior, R. L. (2006). Concentrations of anthocyanins in common foods in the United States and estimation of normal consumption. *Journal of Agricultural and Food Chemistry, 54*(11), 4069–4075.
13. Lutz, M., Hernández, J., & Henriquez, C. (2015). Phenolic content and antioxidant capacity in fresh and dry fruits and vegetables grown in Chile. *CyTA Journal of Food, 13*(4), 541–547.
14. Brønnum Hansen, K., & Flink, J. (1986). Anthocyanin colorants from Elderberry (*Sambucus nigra* L.) IV. Further studies on production of liquid extracts, concentrates and freeze dried powders. *International Journal of Food Science and Technology, 21*(2), 605–614.
15. Friedman, M. (1997). Chemistry, biochemistry, and dietary role of potato polyphenols. A review. *Journal of Agricultural and Food Chemistry, 45*(5), 1523–1540.
16. Pouget, M. P., Lejeune, B., Vennat, B., & Pourrat, A. (1990). Extraction, analysis and study of the stability of *Hibiscus anthocyanins. Lebensmittel–Wissenschaft and Technology, 23*(2), 103–105.
17. Abdulnabi, A., Abushita, E. A., Hebshi, H. G. D., & Biacs, P. A. (1997). Determination of antioxidant vitamins in tomatoes. *Food Chemistry, 60*(2), 207–212.
18. Rosa, J. (1973). Stability of anthocyanin pigment concentrates obtained from black current press cake. II. Studies on the rate of pigment destruction during storage. *Prace Instytut6w i Laboratoridw Badawczych Przemyslu Spolywczego, 23*(3), 447–462.
19. Gil, M. I., Ferreres, F., & Tomas-Barberan, F. A. (1999). Effect of postharvest storage and processing on the antioxidant constituents (flavonoids and vitamin C) of fresh-cut spinach. *Journal of Agricultural and Food Chemistry, 47*(6), 2213–2217.
20. Nakatani, N., Fukuda, H., & Fuwa, H. (1979). Major anthocyanin of Bolivian purple corn (*Zea mays* L.). *Agri Biol Chem, 43*(2), 389–391.
21. Márkus, F., Daood, H. G., Kapitány, J., & Biacs, P. A. (1999). Change in the carotenoid and antioxidant content of spice red pepper (paprika) as a function of ripening and some technological factors. *Journal of Agricultural and Food Chemistry, 47*(1), 100–107.
22. Li, R., Wang, P., Qing-qi, G., & Zhen-yu, W. (2011). Anthocyanin composition and content of the *Vaccinium uliginosum* berry. *Food Chemistry, 125*(1), 116–120.
23. Yang, Z., Yonbin, H., Zhenxin, G., Gongjian, F., & Zhigang, C. (2008). Thermal degradation kinetics of aqueous anthocyanins and visual color of purple corn (*Zea mays* L.) cob. *Innovative Food Science and Emerging, 9*(3), 341–347.
24. Bordonaba, J. G., Crespo, P., & Terry, L. A. (2011). A new acetonitrile-free mobile phase for HPLC-DAD determination of individual anthocyanins in blackcurrant and strawberry fruits: A comparison and validation study. *Food Chemistry, 129*(3), 1265–1273.

25. Patil, G., Madhusudhan, M. C., Ravindra, B. B., & Raghavarao, K. S. M. S. (2009). Extraction, dealcoholization and concentration of anthocyanin from red radish. *Chemical Engineering and Processing, 48*(1), 364–369.
26. Du, Q., Zheng, J., & Xu, Y. (2008). Composition of anthocyanins in mulberry and their antioxidant activity. *Journal of Food Composition and Analysis, 21*(5), 390–395.
27. Arapitsas, P., & Turner, C. (2008). Pressurized solvent extraction and monolithic column-HPLC/DAD analysis of anthocyanins in red cabbage. *Talanta, 74*(5), 1218–1223.
28. McDougall, G. J., Fyffe, S., Dobson, P., & Stewart, D. (2007). Anthocyanins from red cabbage-stability to simulated gastrointestinal digestion. *Phytochemistry, 68*(9), 1285–1294.
29. Nelson, H., Hurtado, A. L., Morales, M., González-Miret, L., Escudero-Gilete, M. L., & Francisco, J. H. (2009). Colour, pH stability and antioxidant activity of anthocyanin rutinosides isolated from tamarillo fruit (*Solanum betaceum* Cav.). *Food Chemistry, 117*(1), 88–93.
30. Nayak, C. A., Rastogi, N. K., & Raghavarao, K. (2010). Bioactive constituents present in *Garcinia indica* Choisy and its potential food applications: A review. *International Journal of Food Properties, 13*(3), 441–453.
31. Ara Kirakosyan, E., Mitchell, S., Kathleen, R., Noon, D. E., Urcuyo, L., Peter, B., et al. (2010). Interactions of antioxidants isolated from tart cherry (*Prunus cerasus*) fruits. *Food Chemistry, 122*(1), 78–83.
32. Ara Kirakosyan, E., Mitchell, S., Kathleen, R., Noon, D. E., Urcuyo, L., Peter, B., et al. (2010). Interactions of antioxidants isolated from tart cherry (*Prunus cerasus*) fruits. *Food Chemistry, 122*(1), 78–83.
33. Markakis, P. (2012). *Anthocyanins as food colors*. New York: Elsevier.
34. Ubi, B. E., Honda, C., Bessho, H., Kondo, S., Wada, M., Kobayashi, S., et al. (2006). Expression analysis of anthocyanin biosynthetic genes in apple skin: Effect of UV-B and temperature. *Plant Science, 170*(3), 571–578.
35. Sapers, G., Taffer, I., & Ross, L. (1981). Functional properties of a food colorant prepared from red cabbage. *Journal of Food Science, 46*(1), 105–109.
36. Chiriboga, C., & Francis, F. (1970). Anthocyanin recovery system from cranberry pomace. *Proceedings of the American Society for Horticultural Science, 95*(2), 233–236.
37. Turfan, O., Türkyılmaz, M., Yemiş, O., & Özkan, M. (2011). Anthocyanin and colour changes during processing of pomegranate (*Punica granatum* L., cv. Hicaznar) juice from sacs and whole fruit. *Food Chemistry, 129*(4), 1644–1651.
38. Huang, Z., Baowu, W., Paul, W., & Ralphenia, D. P. (2009). Identification of anthocyanins in muscadine grapes with HPLC-ESI-MS. *LWT – Food Science and Technology, 42*(4), 819–824.
39. Kraemer-Schafhalter, A., Fuchs, H., & Pfannhauser, W. (1996). Anthocyanins of aronia melanocarpa: Analysis, stability, changes during treatment and storage. In: *Proceedings of the symposium on Polyphenols and anthocyanins as food colorants and antioxidants, Vienna,* 15 *Nov* 1996.
40. Delgado-Vargas, F., Jiménez, A., & Paredes-López, O. (2000). Natural pigments: Carotenoids, anthocyanins, and betalains—Characteristics, biosynthesis, processing, and stability. *Critical Reviews in Food Science and Nutrition, 40*(3), 173–289.
41. Kruger, M. J., Neil, D., Myburgh, K. H., & Sandrine, L. (2014). Proanthocyanidins, anthocyanins and cardiovascular diseases. *Food Research International, 59*, 41–52.
42. Khanal, R. C., Howard, L. R., & Prior, R. L. (2010). Effect of heating on the stability of grape and blueberry pomace procyanidins and total anthocyanins. *Food Research International, 43*(5), 1464–1469.
43. Zhenchang, L., Benhong, W., Peige, F., Chunxiang, Y., Wei, D., Xianbo, Z., et al. (2008). Anthocyanin composition and content in grape berry skin in *Vitis germplasm. Food Chemistry, 111*(4), 837–844.
44. Li, H., Deng, Z., Liu, R., Young, J. C., Zhu, H., Loewen, S., et al. (2011). Characterization of phytochemicals and antioxidant activities of a purple tomato (*Solanum lycopersicum* L.). *Journal of Agricultural and Food Chemistry, 59*(21), 11803–11811.
45. Li, H., Zeyuan, D., Honghui, Z., Chanli, H., Ronghua, L. J., Christopher, Y., et al. (2012). Highly pigmented vegetables: Anthocyanin compositions and their role in antioxidant activities. *Food Research International, 46*(1), 250–259.

46. Ferreres, F., Gil, M. I., & Tomás-Barberán, F. A. (1996). Anthocyanins and flavonoids from shredded red onion and changes during storage in perforated films. *Food Research International, 29*(3), 389–395.
47. Anderson, O., & Jordheim, M. (2006). The anthocyanins. In O. M. Anderson & K. R. Markhand (Eds.), *Flavonoids chemistry, biochemistry and application*. Boca Raton: CRC.
48. Rein, M. (2005). *Copigmentation reactions and color stability of berry anthocyanins*. Helsinki: University of Helsinki.
49. Clifford, M. N. (2000). Anthocyanins-nature, occurrence and dietary burden. *Journal of the Science of Food and Agriculture, 80*(7), 1063–1072.
50. Dey, P. M., & Harborne, J. B. (1997). *Plant biochemistry*. New York: Academic.
51. Andersen, O. M., Jordheim, M., Byamukama, R., Mbabazi, A., Ogweng, G., Skaar, I., et al. (2010). Anthocyanins with unusual furanose sugar (apiose) from leaves of *Synadenium grantii* (Euphorbiaceae). *Phytochemistry, 71*(13), 1558–1563.
52. Tatsuzawa, F., Norio, S., Hiroko, S., Masato, Y., Tomohisa, Y., Koichi, S., et al. (2004). Acylated anthocyanins in the flowers of Vanda (Orchidaceae). *Biochemical Systematics and Ecology, 32*(7), 651–664.
53. Cameira dos Santos, P. J., Jean-Marc, B., Véronique, C., & Michel, M. (1996). Detection and partial characterisation of new anthocyanin derived pigments in wine. *Journal of the Science of Food and Agriculture, 70*(2), 204–208.
54. Glenda, A., Macz-Pop, J. C., Rivas-Gonzalo, J., Pérez-Alonso, J., & Ana, M. G. P. (2006). Natural occurrence of free anthocyanin aglycones in beans (*Phaseolus vulgaris* L.). *Food Chemistry, 94*(3), 448–456.
55. João, P., Sandra, L., Ana, M. G. P., Nuno, M. C., Santos, B., Artur, M. S. S., et al. (2005). Isolation and structural characterization of new anthocyanin-alkyl-catechin pigments. *Food Chemistry, 90*(1), 81–87.
56. Patras, A., Nigel, P. B., Colm, O. D., & Tiwari, B. K. (2010). Effect of thermal processing on anthocyanin stability in foods; mechanisms and kinetics of degradation. *Trends in Food Science and Technology, 21*(1), 3–11.
57. Fan-Chiang, H. J., & Wrolstad, R. E. (2005). Anthocyanin pigment composition of blackberries. *Journal of Food Science, 70*(3), C198–C202.
58. Kuroda, C., & WADA, M. (1933). The colouring matter of eggplant (Nasu 1). *Proceedings of the Imperial Academy, 9*(2), 51–52.
59. Solomon, A., Golubowicz, S., Yablowicz, Z., Grossman, S., Bergman, M., Gottlieb, H. E., et al. (2006). Antioxidant activities and anthocyanin content of fresh fruits of common fig (*Ficus carica* L.). *Journal of Agricultural and Food Chemistry, 54*(20), 7717–7723.
60. Dueñas, M., José, J. P., Santos-Buelga, C., & Escribano-Bailón, T. (2008). Anthocyanin composition in fig (*Ficus carica* L.). *Journal of Food Composition and Analysis, 21*(2), 107–115.
61. Warner, L. M. (Ed.). (2015). *Handbook of anthocyanins: Food sources, chemical applications and health benefits*. New York: Nova Science.
62. Takeoka, G. R., Dao, L. T., Tamura, H., & Harden, L. A. (2005). Delphinidin 3-O-(2-O-β-D-glucopyranosyl-α-L-arabinopyranoside): A novel anthocyanin identified in beluga black lentils. *Journal of Agricultural and Food Chemistry, 53*(12), 4932–4937.
63. Montoro, P., Tuberoso, C. I., Perrone, A., Piacente, S., Cabras, P., & Pizza, C. (2006). Characterisation by liquid chromatography-electrospray tandem mass spectrometry of anthocyanins in extracts of *Myrtus communis* L. berries used for the preparation of myrtle liqueur. *Journal of Chromatography A, 1112*(1), 232–240.
64. Aparicio-Ruiz, R. N., Gandul-Rojas, B., & Roca, M. (2009). Pigment profile in non-Spanish olive varieties (*Olea europaea* L. Var. Coratina, Frantoio, and Koroneiki). *Journal of Agricultural and Food Chemistry, 57*(22), 10831–10836.
65. Grigoras, C. G., Emilie, D., Sandrine, Z., & Claire, E. (2012). Sweet cherries anthocyanins: An environmental friendly extraction and purification method. *Separation and Purification Technology, 100*, 51–58.
66. Usenik, V., Fabčič, J., & Štampar, F. (2008). Sugars, organic acids, phenolic composition and antioxidant activity of sweet cherry (*Prunus avium* L.). *Food Chemistry, 107*(1), 185–192.

67. Kim, D. O., Chun, O. K., Kim, Y. J., Moon, H. Y., & Lee, C. Y. (2003). Quantification of polyphenolics and their antioxidant capacity in fresh plums. *Journal of Agricultural and Food Chemistry, 51*(22), 6509–6515.
68. Chun, O. K., Kim, D.-O., & Lee, C. Y. (2003). Superoxide radical scavenging activity of the major polyphenols in fresh plums. *Journal of Agricultural and Food Chemistry, 51*(27), 8067–8072.
69. Fossen, T., Andersen, O. M., Dag, O., Ovstedal, A. T., & Pedersen, A. R. (1996). Characteristic anthocyanin pattern from onions and other *Allium* spp. *Journal of Food Science, 61*(4), 703–706.
70. Donner, H., Gao, L., & Mazza, G. (1997). Separation and characterization of simple and malonylated anthocyanins in red onions, *Allium cepa* L. *Food Research International, 30*(8), 637–643.
71. Jing, P., Shu-Juan, Z., Si-Yu, R., Zhuo-Hong, X., Ying, D., & Liangli, Y. (2012). Anthocyanin and glucosinolate occurrences in the roots of Chinese red radish (*Raphanus sativus* L.), and their stability to heat and pH. *Food Chemistry, 133*(4), 1569–1576.
72. Carazzone, C., Mascherpa, D., Gazzani, G., & Papetti, A. (2013). Identification of phenolic constituents in red chicory salads (*Cichorium intybus*) by high-performance liquid chromatography with diode array detection and electrospray ionisation tandem mass spectrometry. *Food Chemistry, 138*(2), 1062–1071.
73. Mulabagal, V., Haibo, W., Mathieu, N., & Muraleedharan, G. N. (2009). Characterization and quantification of health beneficial anthocyanins in leaf chicory (*Cichorium intybus*) varieties. *European Food Research and Technology, 230*(1), 47–53.
74. Mena, P., Calani, L., Dall'Asta, C., Galaverna, G., García-Viguera, C., Bruni, R., et al. (2012). Rapid and comprehensive evaluation of (poly) phenolic compounds in pomegranate (*Punica granatum* L.) juice by UHPLC-MSn. *Molecules, 17*(12), 14821–14840.
75. Zhao, X., Zhaohe, Y., Fang, Y., Yanlei, Y., & Lijuan, F. (2013). Characterization and evaluation of major anthocyanins in pomegranate (*Punica granatum* L.) peel of different cultivars and their development phases. *European Food Research and Technology, 236*(1), 109–117.

Chapter 4
Anthocyanins as Natural Colors

4.1 Introduction

Colors play a very important but often under-estimated role in our lives. Since ancient times natural colors derived from plants, lichens and insects have been used as sources for natural colors, suitable for multiple applications including textiles, paints, works of art, food, clothing and cosmetics. Colors fascinate human and animals as they are one of the first characteristics observed by senses and helps in determining the acceptability of food and play a crucial role in the plant pollination process by animals.

Since, a decent outlook is crucial to fetch high market price of food item, natural or synthetic color is added to the food. Natural pigments from plant or animal origin are believed to be more safe and cost-effective than synthetic ones. Synthetic dyes have been suspected to affect adversely central nervous system both behaviorally and neurologically. Colors from biological origin are called 'biocolors'. The success of any color, natural or synthetic, depends upon its consumer-acceptance, approval by regulatory bodies and the size of the investment required in bringing the color to store-shelves. The use of natural colors is an influential tool for selling a product in market. The food as such or in packed form is judged by its color or even by color of package. It has led to claims appearing on colored food products and commodities like, "no synthetic added" or "all natural".

There are many pigmented classes of phytochemicals, such as carotenoids, flavonoids and anthocyanins. The later class has been focus of attention as these are safe compared to artificial colors like Sudan red III and other synthetic colors. Anthocyanins like other bio-colors have a high market value with prices variations from US $1250 to $2000 per kg. No permission is required for work on these pigments as these occur naturally.

Chlorophyll, carotenoids, and anthocyanins are the most abundant natural occurring pigments in plant tissues. We recognize anthocyanins indirectly because of the splendid red to red-brown to purple autumn colors of the leaves of trees. In combi-

M. Riaz et al., *Anthocyanins and Human Health*, SpringerBriefs in Food, Health, and Nutrition, DOI 10.1007/978-3-319-26456-1_4

nation with the yellow red carotenoids, they give leaves, flowers and fruits a wide spectrum of colors. The anthocyanins are now exclusively used as food colors but in older times they were used in paints for precious manuscripts. They are now prepared from the peels of grapes (enocyanin) as a side product of the grape harvest and winemaking. Red cabbage is also a good source of anthocyanins. The rainbow of colors of anthocyanins is due to conjugated bonds present in their structures. Most of the flower colors are flavonoids in nature and red or purple is associated with anthocyanins alone or as co-pigment responsible for other related colors. The pigmentation of flower attract insects and contribute in its propagation, another important role of anthocyanins for plants is protection from external plant enemies e.g. cyanidin and peonidin glucosides inhibited the growth of *Xanthomonas oryzae* [1]. Their importance may not be overlooked because they have been sub-grouped as color of current EU approved additives with E163, E-Number by EU food standard agency [2]. Safety and water solubility of anthocyanins allow their integration into aqueous foodstuff without unwanted negative effects [3].

Like synthetic color, pure single natural color is hard to obtain thus extract from the concerned natural source is used as coloring agent for example the most commonly anthocyanins enriched extract used are grape peel and black currant. This natural color varies with the stage and season of the plant, leaves when young have different color than aged leaves and same is for fruit and flowers. The reason for color variation is thought to be the intra and intermolecular bioconversion of anthocyanins or derivatives with age and environment to its variant analog moieties that have specific colors. For example the interconversion of aromatic polyacylated anthocyanins is responsible for bluing effect and stability of flower color. Taking the advantage of the advancement in plant biotechnology the cultivar of desired color may be produced by incorporating the gene of interest. Investigation on this side is required to produce natural color in greater in quantity and quality.

As mentioned above, the color of anthocyanins is due to the conjugated double bonds carrying a positive charge on the heterocyclic oxygen ring under acidic conditions. Various factors like pH, co-pigmentation and interactions with metal ions lead to variation of colors in anthocyanins by affecting the conjugated double bond system present in basic skeleton of anthocyanins. At pH near to 1 the free flavylium cation is red while at higher pH, the color changes to purple and blue as it goes in transition to the quinoidal base and subsequently to ionized quinoidal base. At higher pH, the colorless carbinol pseudo-base is formed which ultimately changes to chalcone molecule having a pale yellow color. Substitution of hydroxyl or methyl group also affect the color of anthocyanins (Figs. 4.1 and 4.2) for example increasing the number of hydroxyl increase bluishness, while increasing the number of methyl group intensify redness [1].

Anthocynanin color depends upon number of OH groups attached especially on B ring. Addition of OH groups shifts their color from orange to violet. Similarly glycosylation of anthocyanidin leads to extra red color of anthocyanins while presence of aromatic or aliphatic acyl moieties does not change color but affects anthocyanin solubility as well as stability. Literature indicates that more than 650 anthocyanins exist in nature. In 1993, over 250 different anthocyanins have been

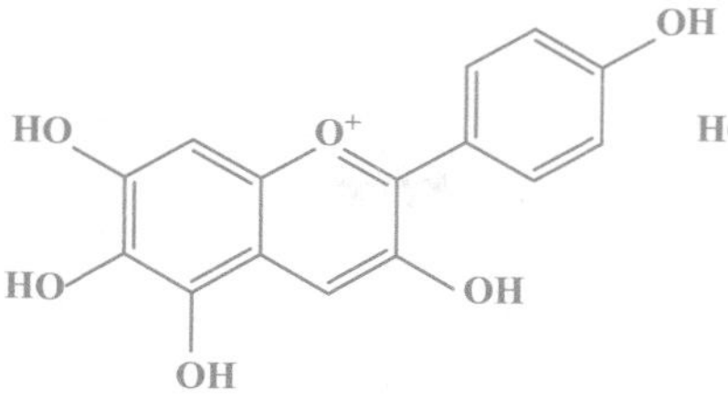

Aurantinidin
(Orange)

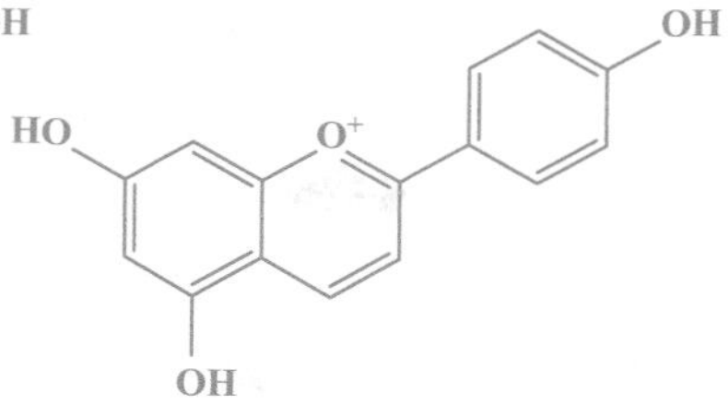

Apigeninidin
(Orange)

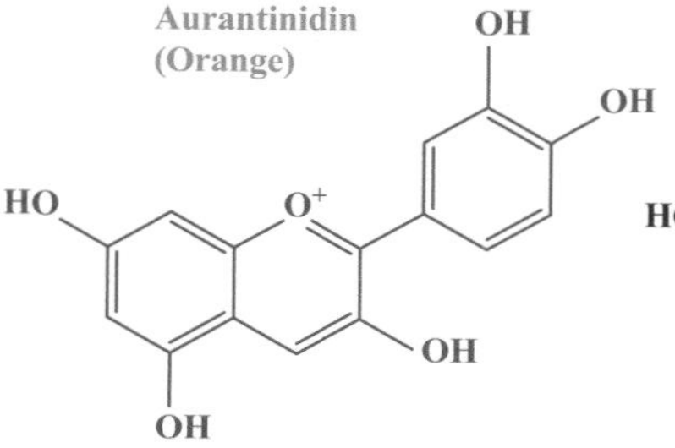

Cyanidin
(Magenta and Crimson)

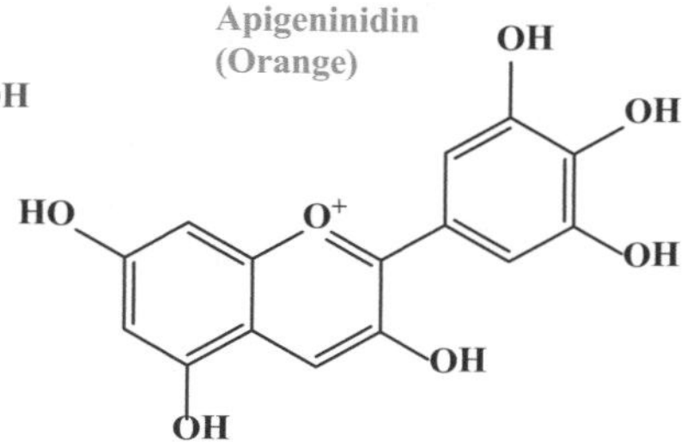

Delphinidin
(Purple, mauve and blue)

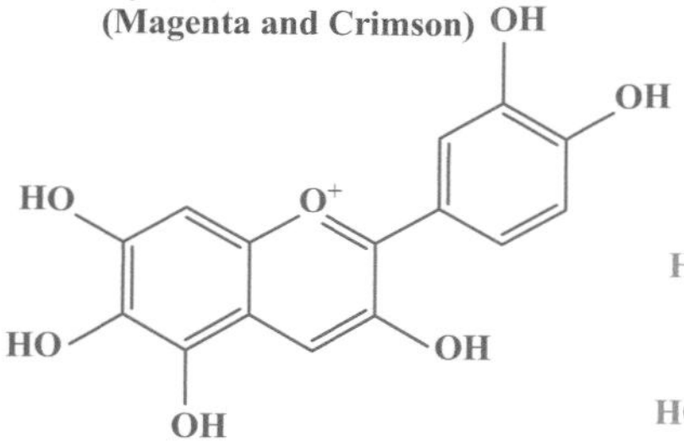

6-Hydroxycyanidin
(Red)

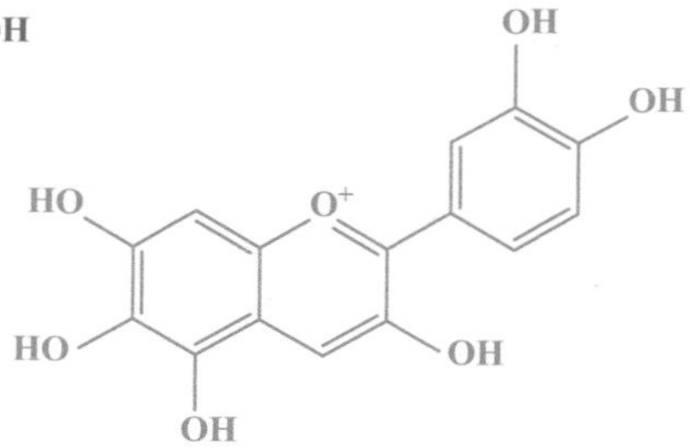

Luteolinidin
(Orange)

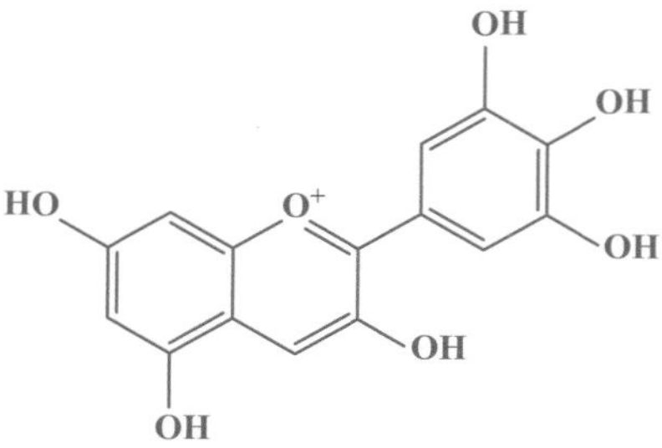

Triacetidin
(Red)

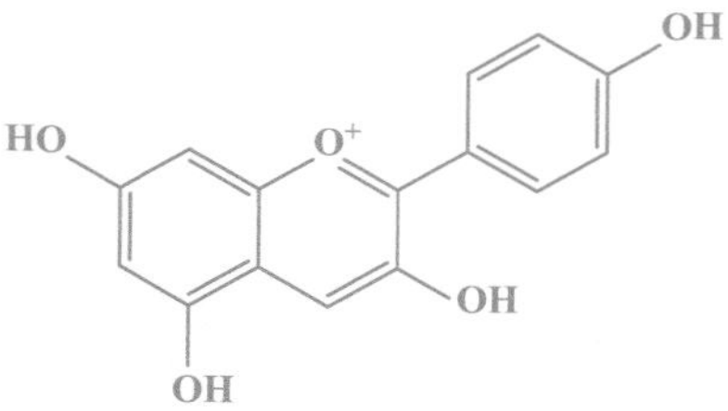

Pelargonidin
(Orange, salmon)

Fig. 4.1 Natural OH substituent's anthocyanidins and change in color [1]

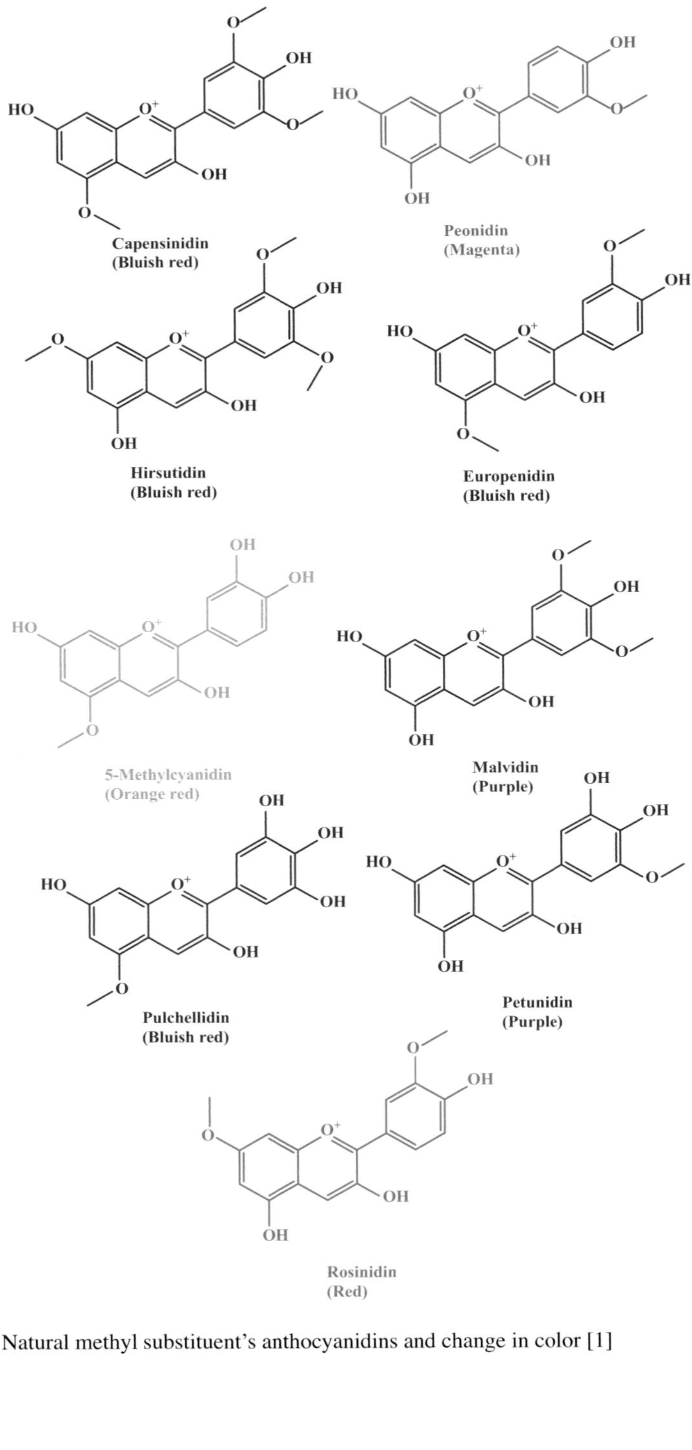

Capensinidin
(Bluish red)

Peonidin
(Magenta)

Hirsutidin
(Bluish red)

Europenidin
(Bluish red)

5-Methylcyanidin
(Orange red)

Malvidin
(Purple)

Pulchellidin
(Bluish red)

Petunidin
(Purple)

Rosinidin
(Red)

Fig. 4.2 Natural methyl substituent's anthocyanidins and change in color [1]

isolated from plants [4–7] and 500 anthocyanin structures reported by the year 2000 [8]. The six common anthocyanidins yield greater than 540 anthocyanins in nature [9], due to acylation of sugar residues with organic-acids and structural differences of glycosidic substitution at the 3 and 5 positions [10]. Andersen and Jordheim, in 2006 reported the number 500 [10] and the number increased to 635 in 2008 [11].

Anthocynains are exclusively used as red color for food stuff although sometime betalains are also used. The chief reason of this use in food matrixes is requirement of marinating the acidic pH (less than 3.5) to obtain the required red color. Grape extracts are more stable towards pH variations so they are widely used as commercial sources of anthocyanins. Further grapes are available globally since they are a major fruit in many countries.

The cheap source of anthocyanin is by-products of some industrial practices like grape skin extracts. Therefore more focus is being given to anthocyanin-rich waste by-products like purple corncobs and banana bracts. The use of these by-products is however very challenging. Other ingredients can be added. Designing an accurate recipe of different reagents to increase their color stability as well suitable processing and storage time can lead to stable colors.

Although being obtained from plants, there are many challenges. For example the supply is subjected to long cultivation times, climatic and seasonal variations, disease risks, increasing cost of agriculture as well as declining availability of cultivable land all making their availability highly challenging. New feasible candidate to be investigated as new natural pigment must be readily available, economical and high yield plant and with satisfactory tinctorial strength. The great activity in this field is supported by data from a survey by Frost and Sullivan in 2002, who predict an expansion of the European color market by 1 % per year till 2008, whereas coloring foodstuff is estimated to grow even by 10–15 % in the same time range [12].

4.2 Use of Anthocyanin-Based Colorants

Anthocyanins are being used by humans since beginning of civilization for decorative purposes. Egyptians used wine as source of food color to enhance the color of candies approximately 1500 BC. The oldest anthocyanin used and marketed is enocynain obtained from pomace of red grape in Italy since 1879. According to the USA regulations, "color additives are defined as any dye, pigment or substance capable of imparting color when added or applied to a food, drug or cosmetic (21CFR70.3). ANs are permitted as food colorants in the USA under the category of fruit (21CFR73.250) or vegetable (21CFR73.260) juice color." Under these categories only suitable for eating part of plants and their water extracts are permitted for use.

Anthocyanins themselves and any of anthocyanin-derived colorant are recognized as a natural colorant (labeled as E163) by the European Union in *Codex Alimentarius* system. In USA, the Food and Drug Administration Authority (FDA) has cataloged anthocyanin under list of natural colors that do not require certifications, i.e. without FD&C numbers. Out of these exempted 26 colorants approved by

FDA, 4 are based on anthocyanin pigments. Anthocyanins are a potent alternative to FD&C Red No. 40 which is a synthetic dye with highest consumption in US. Various food plants have been considered as marketable source of anthocyanin-based colorants, however their use has been limited by pigment stability, availability of raw material and economic considerations [3].

4.3 Acylated Anthocyanins as Colorants for the Food Industry

Anthocyanins are potential food color substitute however the major demerit is poor stability, however acylation perk up color and pigment stability. These acylated anthocyanins have variety of shades that are dependent on pH of food matrix and structure changes in anthocyanins. For example acylated pelargonidin derivatives obtained from radish and potato at acidic pH match allura red and acylated cyanidin from black carrots or cabbage pH dependent hues ranging from deep red to purple. More than 65 % of reported anthocyanins whose structures have been properly identified are acylated. Anthocyanins diversity greatly depends on number, nature and linkage position of acyl groups.

Acyalted derivatives of anthocyanins are stable and have wide color range so scientists are in search of acylated anthocyanin-rich material (Table 4.1) for example red cabbage in Polish food industry [13, 14] has color like blueberries at pH 4.2–4.5 and usually the extract of red cabbage is available in a liquid or sprayed-dried water-soluble form [3]. Another exceptional source of acylated anthocyanins is black carrot (*Daucus carota* L) consumed in Afghanistan, Egypt, Pakistan, Spain, and Turkey [3]. It is well-known now that poly acylated anthocyanins are more stable while simple anthocyanins quickly drop their colors by hydration at 2-position of anthocyanidin nucleus.

Another acceptable source of stable acylated anthocyanins is red potatoes [15, 16]. Extracts of acylated anthocyanins can be applied for foods having a low pH level, including jellies, sauces, conserves, confectionery and bakery items and soft drinks [17]. Some other potential applications of acylated anthocyanins are neutral or slightly alkaline products, such as powdered and ready-to-eat desserts, milk-drinks, ice cream and panned products [3, 17]. Table 4.1 shows sources of some acylated anthocyanins, their properties and their use in food industry.

Anthocyanins react very rapidly with metals forming stable complexes with iron, copper and tin, these complexes can be used as potential coloring agents due to their higher stability. Acylation decrease water solubility while glycosylation increases it. Hydroxylation (with more –OH groups) produces blue colors while methoxylation (with more methoxyl groups) produce red colors. The number and type of glycosidic units at C-3 don't change color significantly while additional glycosylation at C-5 induces a slight shift to red-purple.

Table 4.1 Sources of acylated anthocyanins, their properties and their uses in food industry

Source	Properties	Types of anthocyanins	Trade name	Company
Maqui berry (*Aristotelia chilensis* L)	Excellent ROS/RNS scavenger, ↓ oxidative stress, blood glucoe, platelet activity	Derivatives of delphinidin 3,5-O-diglucoside, delphinidin 3-O-samb, 5-O-glucoside, delphinidin 3-O-glucoside	Delphinol®	Maqui New Life
Red cabbage (*Brassica oleracea* L)	Expensive, produces different shades at different pH, have pleasant taste, stability to heat and light, low in polyphenols reducing the risk of hazing with proteins, available all year round and have potential health benefits when included into the diet	15 Anthocyanins (mainly diacylated) being derivatives of cyanidin-3-diglucoside, -5-glucoside acylated with sinapic, ferulic and/or *p*-coumaric acids	Magento™	Overseal Foods Ltd
Black carrot (*Daucus carota* L)	Provide an excellent bright strawberry red shade in acidic products, exhibit mauve to blue tones under neutral pH, vegetarian and kosher alternative to carmine, has low levels of polyphenols and improved stability to heat, light and SO_2	Derivatives of cyanidin-3-rutynoside--glucoside-galactoside acylated with one cinnamic acid (*p*- -coumaric, ferulic or sinapic)	Exberry® Carantho®	GNT Group, Overseal Foods Ltd
Red radish (*Raphanus sativus* L).	Acylated pelargonidin derivatives from red radish give red color to maraschino cherries very close to that of synthetic colorant FD&C Red # 40 at pH 3.5	This vegetable contains 12 acylated anthocyanins (8 diacylated)	ColorPure™	RFI
Red potatoes (*Solanum tuberosum*)	Possible natural alternative pigment source to synthetic colorant FD&C Red # 40	Monoacylated anthocyanins	Not available	Nil
Purple corn (*Zea mays*) and oxalis (*Oxalis triangularis*)	Under consideration as potential food color sources	Not yet known	Not available	Nil

4.4 Conclusions

Pigments act as fingerprint of a food commodity determining its overall quality. Anthocyanin-rich extracts obtained from purple corn, blackcurrant, grape skin and red cabbage are now approved as natural colorants in New Zealand, Australia and Europe and are commercially used in food industry. However, anthocyanins as natural colors have their own limitations. Difficulties in collection of plant materials, lack of standardization, extraction procedure from plants material and low color value makes them less focused as substituent to synthetic colors. If somehow obtained and used, they face problem of instability due to temperature, oxygen, light, pH and enzymes. Further the color imparted is fugitive sometimes and a mordant needs to be added to increase colors. However it should not be a disappointing. Their stability can be increased by adding stabilizers like dextrin additives obtained from tart cherries. Further research should be conducted for identification, characterization, documentation, assessment and potential utilization from un-prospected plant species. The large scale cultivation from these species and increasing anthocyanin contents in existing species by genetic engineering will lead to availability of sufficient amounts for industrial extraction, improvement of quality and quantity of anthocyanins and enhance their stability during processing and storage.

References

1. Delgado-Vargas, F., Jiménez, A., & Paredes-López, O. (2000). Natural pigments: Carotenoids, anthocyanins, and betalains—Characteristics, biosynthesis, processing, and stability. *Critical Reviews in Food Science and Nutrition, 40*(3), 173–289.
2. Current EU approved additives and their E numbers. (2014). *EU Food Standard Agency*. http://www.food.gov.uk/science/additives/enumberlist.
3. Giusti, M. M., & Wrolstad, R. E. (2003). Acylated anthocyanins from edible sources and their applications in food systems. *Biochemical Engineering Journal, 14*(3), 217–225.
4. Mazza, G., & Miniati, E. (1993). *Anthocyanins in fruits, vegetables, and grains*. Boca Raton: CRC.
5. Harborne, J. B., & Williams, C. A. (2001). Anthocyanins and other flavonoids. *Natural Product Reports, 18*(3), 310–333.
6. Kong, J. M., Chia, L. S., Goh, N. K., Chia, T. F., & Brouillard, R. (2003). Analysis and biological activities of anthocyanins. *Phytochemistry, 64*(5), 923–933.
7. Strack, D., Steglich, W., & Wray, V. (1993). Betalains. *Methods in Plant Biochemistry, 8*, 421–450.
8. Pietta, P.-G. (2000). Flavonoids as antioxidants. *Journal of Natural Products, 63*(7), 1035–1042.
9. Andersen, Ø. M., Fossen, T., Torskangerpoll, K., Fossen, A., & Hauge, U. (2004). Anthocyanin from strawberry (*Fragaria ananassa*) with the novel aglycone, 5-carboxypyranopelargonidin. *Phytochemistry, 65*(4), 405–410.
10. Wrolstad, R. E., Durst, R. W., & Lee, J. (2005). Tracking color and pigment changes in anthocyanin products. *Trends in Food Science and Technology, 16*(9), 423–428.

11. Anderson, Ø., & Jordheim, M. (2008). *Anthocyanins: Food applications*. In *Proceedings of 5th International Congress Pigments in Food: For Quality and Health, University Helsinki Helsinki, Finland*.
12. Stintzing, F. C., & Carle, R. (2004). Functional properties of anthocyanins and betalains in plants, food, and in human nutrition. *Trends in Food Science and Technology, 15*(1), 19–38.
13. Baublis, A., Spomer, A., & Berber Jiménez, M. D. (1994). Anthocyanin pigments: Comparison of extract stability. *Journal of Food Science, 59*(6), 1219–1221.
14. Giusti, M. M., Luis, E. R., Donald, G., & Ronald, E. W. (1999). Electrospray and tandem mass spectroscopy as tools for anthocyanin characterization. *Journal of Agricultural and Food Chemistry, 47*(11), 4657–4664.
15. Fossen, T., & Andersen, Ø. M. (2003). Anthocyanins from red onion, (Allium cepa) with novel aglycone. *Phytochemistry, 62*(8), 1217–1220.
16. Terahara, N., Konczak, I., Ono, H., Yoshimoto, M., & Yamakawa, O. (2004). Characterization of acylated anthocyanins in callus induced from storage root of purple-fleshed sweet potato, *Ipomoea batatas* L. *BioMed Research International, 2004*(5), 279–286.
17. Naturex. (2014). *Ultimate botanical benifits*. Cited September 2014.

Chapter 5
Anthocyanins Absorption and Metabolism

5.1 Introduction

Biologically-active compounds, after ingestion, interact during metabolism to modify biological responses that can either strengthen or restrict their efficacy and potency. Regarding anthocyanins, as discussed before in Chaps. 2 and 3, no member of animal kingdom can synthesize these molecules and they have to rely on plants. Although humans are not colored by anthocyanins, low to significant contents of anthocyanins are observed in human blood, tissues and organs. Demographical studies indicate that most of the world population lives in poverty and do not have luxury to afford such dietary recommendations. The efficiency of food in digestion and bioavailability is the chief factor, i.e. the amount of anthocyanins released, solubilized, absorbed, transported and metabolized. It is once the anthocyanins are in human body that the major uncertainty begins. The ratio of total digested anthocyanins to bioavailable anthocyanins must always be considered. The outcomes are highly inconsistent and sometimes contradictory. Due to variations of experimental-designs, analytical procedures and species analyzed, the statistical figures reported and arithmetical comparisons between studies are not precise.

Stability and functionality of anthocyanins in human body depends on their content, type, and location in food matrix and presence of other bioactive compounds in fruits and vegetables. Their instability depends directly on number of –OH groups and indirectly on number of $–OCH_3$ groups. Further glycosylation also affects stability as diglucoides are more stable than monoglucosides. Being polar in nature, anthocyanins are more soluble in polar solvents as compared to non-polar solvents. Further food processing can lead to qualitative and quantitative changes since anthocyanins are susceptible to oxidation and isomerization. Their bioefficacy depends upon their bioavailability and bioaccessibility. Studies indicate that bioavailability is variable depending upon various diverse parameters. Anthocyanidins and anthocyanins are the least absorbed phenolics. Therefore, their abundance in term of dietary intake does not necessarily mean their intact and relative metabolites in human body.

M. Riaz et al., *Anthocyanins and Human Health*, SpringerBriefs in Food, Health, and Nutrition, DOI 10.1007/978-3-319-26456-1_5

The main criterion for absorption and resultant exerted bioactivity is liberation from food matrix, i.e. bioaccessibility. Bioaccessibility indicates the amount of anthocyanins available for absorption in gut. Bioaccessibility and bioavailability depends on the physicochemical characteristics of food matrix which affects the efficacy of the enzymatic and physicochemical digestion processes. Actually food micro-structure determines release, transformation and absorption of anthocyanins in digestive tract. Complexation with other food components, attachment to specific organelles and restriction within intact cell walls or entrenchment in food matrix may affect the release of anthocyanins. Consequently anthocyanins bioavailability and bioactivity depends upon food matrix behavior, the presence of promoters and inhibitors of anthocyanins absorption. Food processing disrupts the food matrix which can affect their bioavailability. While the analytical methods themselves are subject to various sources of errors, the main problem with anthocyanins analysis lies in their inherent instability.

5.2 Daily Intake

Currently, we are watching a worldwide sharp increase in research activity as well as in dietary supplement containing anthocyanins. Despite the fact that the need for anthocyanins is different and depends on income level, information of anthocyanins content and composition is necessary so that guidelines can be provided on food sources that can provide ample to rich supplies of preferred anthocyanins. Anthocyanins are quickly being incorporated into the mainstream channel of functional foods, dietary supplements and nutraceuticals and progressively being recognized by the public due to the demand for health-promoting diets. Depending upon the dietary habits and socioeconomic conditions, the daily intake of anthocyanins varies from micrograms to hundreds of grams per person. Although individuals in developed countries use more anthocyanins due to awareness of their health-promoting effects created by television commercials and all types of media-campaigns, intake of diets containing anthocyanins is gradually increasing in low-income countries also because of easy commercial and market availability of juices, drinks, slushes, concentrates and extracts of fruits and vegetables containing greater anthocyanins contents. Since anthocyanins are not classified as essential nutrients so their recommended daily intake (RDI) has not been established yet. Daily intake of anthocyanins in USA had previously (1976) been estimated to be 180–215 mg per day per person [1], but a recent study by USDA, based on the evaluation of more than 100 common foods in the USA, daily intake was found to be 12.5 mg per day per person [2]. Interestingly 100 g of berries offer up to 500 mg of anthocyanins [3]. This intake of anthocyanins is higher compared with dietary flavonoids such as genistein and quercetin (estimated at 20–25 mg/day) [4]. The daily intake of anthocyaninsis higher compared than in the USA. According to multinational study across ten European countries report (36,037 individuals in the age of 35–74 years), the mean daily intake of anthocyanidins [5] ranged in men from

19.8 to 64.9 mg, while in women from 18.7 to 44.1 mg. In addition, this study showed high anthocyanidins intake by southern European, non-obese and non-smoker older women [6]. The different values of daily intake are mainly due to different food intake data. Authentic and exact food intake data along with complete food anthocyanin concentration data is necessary for estimation of daily intake.

Proanthocyanidins content of food fluctuates depending upon distribution of monomers (flavan-3-ols), oligomers and polymers (usually N10, flavan-3-ol monomers), and their inter-flavan bonds [7]. The mean daily intake of oligomeric and polymeric proanthocyanidins was estimated to be higher than that of monomeric flavan-3-ols, and twice as high as the combined overall intake of other flavonoids [8]. The total proanthocyanidin intake by American adults range from 96 to 137 mg per day (mg/day) [9],whereas those reported for Spanish 189 mg/day and for Finnish adults was [10] 128 mg/day [11]. Wang et al. (2011) recorded the mean daily intake in American adults to be 95 mg/day [12], the value of intake lies at lower end of amount that was previously seen by Gu et al. [9, 13].

5.3 Anthocyanins Absorption

The preventive and therapeutic actions of anthocyanins are directly proportional to their absorption and metabolism in human body. Many diverse challenges exist in conducting kinetic studies in humans mainly due to their low levels in circulation. Comprehensive information of qualitative and quantitative distribution of anthocyanins in food chain, human blood, tissues and organs is crucial in order to establish a statistically significant correlation between consumption of anthocyanins and their preventive and pharmacological effects. The absorption of anthocyanins depends upon many factors including chemical structure, nature of food matrices, interaction with other micro, macro and phytonutrients, type and extent of food processing and preparation, as well as the nutritional, pathophysiological and genetic factors of individual. The change in absorption of anthocyanins is also associated with its biological action, in general the absorption of the anthocyanins are less, so this point may be used for increasing the biological actions by playing with anthocyanins pharmacokinitics via different ways. However the structural and functional differences should be considered with respect to the source of anthocyanins for example the vegetable anthocyanins are more complex than fruit and thus their hydrolysis rate, bioavailability and metabolism is different.

Various factors can influence the ability to absorb, convert and metabolize the anthocyanins. Due to their nutritive and health promoting effects, interest in factors that influence their bioavailability has intensified. Biological activities of anthocyanins are dependent on its bioavailability, i.e. the amount of anthocyanins reach to systemic circulation and it is usually based on absorption and metabolism. Despite much information, still our knowledge about anthocyanins distribution and availability is limited. Usually acylated anthocyanins are less absorbed than non-acylated ones. The type and numbers of backbone and acyl group are believed to affect their bioavailability in humans. Bioavailability from supplements is usually much better. In mouth, chewing releases anthocyanins from the food matrix. The bioavailability of anthocyanins is very low, with <1 % of the ingested amount reaching the plasma, glucuronidated and methylated anthocyanin metabolites were found twice than their parent (intact) compounds [14]. In human body anthocyanins interact physically and chemically with other phytochemicals which may lead to synergism or interference.

5.3.1 Gastric Absorption

Anthocyanins absorption may start in the stomach and appear in the blood immediately after ingestion. Passamonti et al. (2003) and Talavera et al. (2003) used analogous methods to establish that anthocyanins were proficiently absorbed in the stomach [15, 16]. Passamonti et al. injected grape anthocyanins into the stomach of 19 Wistar male rats and collected blood from both the portal vein and the heart at 6 min. Malvidin-3-glucoside was present in both portal and systemic plasma. Importantly, malvidin-3-glucoside appeared in the plasma within 6 min, presenting an evidence of stomach absorption. Peonidin-3-glucoside, petunidin-3-glucoside, and malvidin-3-acetyl glucoside derivatives were inconsistently detected, perhaps owing to animal variability. Neither anthocyanins aglycones nor conjugated derivatives were detected in the plasma. Talavera et al. [16] (2003) infused anthocyanin standards as well as bilberry and blackberry extract into the stomach of pylorus- and sphincter-ligated rats. Gastric contents sample from gastric vein and blood sample from abdominal aorta were taken 30 min after the administration. HPLC analysis revealed that a high proportion (~25 %) of anthocyanin mono-glycosides, including glucoside and galactoside, was absorbed from the stomach, whereas the rutinoside was poorly absorbed. It was suggested that gastric absorption of anthocyanins involves bilitranslocase (TC 2.A.65.1.1), an organic anion membrane carrier in the gastric mucosa [17].

5.3.2 Absorption in the Small Intestine

The small intestine is the major site for anthocyanins absorption. Absorption of anthocyanins in small intestine of anesthetized rats was estimated by an in situ perfusion method [18] supplemented in physiological buffers. The amount of anthocyanins remaining in the effluent was used to estimate the rate of anthocyanins absorption in the small intestine. Depending on their structures, the absorption rate of supplemented anthocyanins ranged from 22.4 ± 2.0 % (cyanidin-3-glucoside) to 10.7 ± 1.1 % (malvidin-3-glucoside). Such high absorption rates seemed to contradict the very low levels of anthocyanins observed in the blood [19]. However, it has to be noted that these absorption rates were calculated based on the disappeared amount in the effluent, thus they could indicate the portion of anthocyanins being taken up into the small intestine tissue but not necessarily transferred into the blood. Recently, our research group also demonstrated that as high as 7.5 % of the administered black raspberry anthocyanins could be taken up by rat small intestinal tissue, yet very limited amount can be detected in urine [20].

Endogenic β-glucosidases are involved at this stage to release aglycones from anthocyadin-glucosides. Since anthocyanin are highly water soluble and are very large molecules, their absorption cannot take place by passive diffusion. Therefore absorption of anthocyanins should be either hydrolyzed to the aglycone in small intestine by α-rhamnosidase, β-glucuronidase or β-glucosidase or they should utilize an active transport mechanism to transport glycosides across the intestinal wall. Aglycones in free status are more hydrophobic and are smaller than the glycosides and can easily infiltrate the epithelial layer passively. However intact glycosides are absorbed by the small intestine, either by sodium-dependent glucose transporter (SGLT1) or by the inefficient passive diffusion. Recent evidence have suggested that acylated anthocyanins are somewhat bioavailable in the intact form [21], although, likely owing to their increased molecular size, acylated ANs are much less efficiently absorbed than their counterparts without the acylation [22].

Phase II enzymes convert anthocyanins to glucuronids, methylates and sulfates in liver and kidney [23, 24]. These conjugated forms of anthocyanins may be excreted via bile to jejunum and recycled by enterohepatic circulation system in intestine/colon. Figure 5.1 indicates the potential mechanism of anthocyanins (ANs) absorption

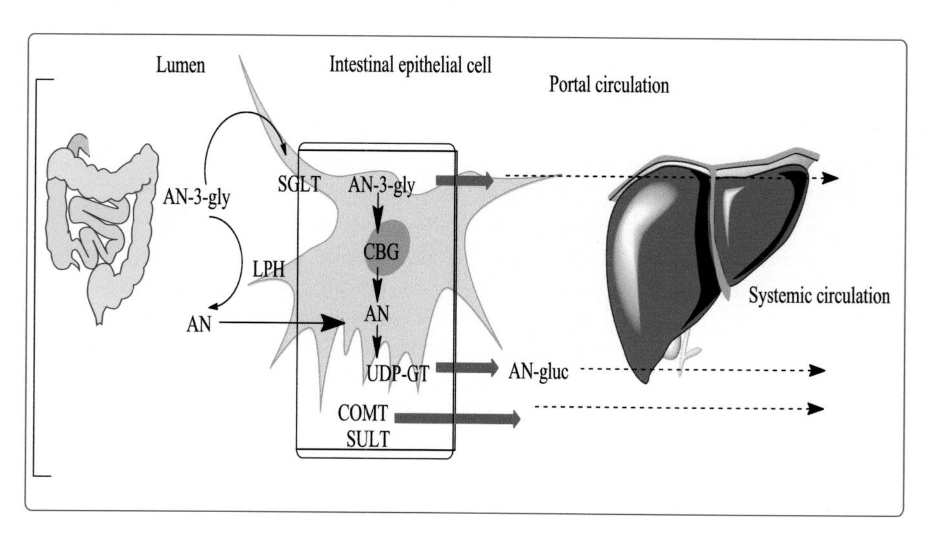

Fig. 5.1 Potential mechanism of anthocyanins (ANs) absorption, sodium-glucose co-transporter, *AN-3-gly* anthocyanins-3-glycoside, *CBG* cystolic B-glucosidase, *LPH* lactate pholorizin hydrolase, *UDP-GT* UDP-glucuronosyl-transferase, *AN-gluc* anthocyanins glucuronide, *COMT* catechol-O-methyl=transferase, *SULT* sulfotransferase [25]

Many studies have confirmed the blood circulation of anthocyanins after fast-track absorption in intact form from stomach and small intestine and their excretion as methylated, sulfo-conjugated or glucueonide derivatives via urinary system [15, 16, 23, 26]. These are one of few anthocyanins that are found in intact forms in blood plasma [27]. In a recent study when anthocynain-rich diet was fed to rats for almost 17 days, their presence was observed in stomach, small intestine, liver, kidney and brain. Total anthocyanin content in brain (blackberry anthocyanins and peonidin 3-*O*-glucoside) was found 0.25±0.05 nmol/g of tissue [14, 28].

5.3.3 Pharmacokinetics

The knowledge of pharmacokinetics is vital to comprehend the effect of daily intake of anthocyanins on health-improvement. There is limited information on complete pharmacokinetic data of anthocyanins. Challenges exist in understanding kinetics of anthocyanins in biological systems due to their low levels in human circulation making it hard to observe their absorption and metabolism. Despite this limitation, considerable work has been carried out in last decade in assessing the dynamics and kinetics of anthocyanins in humans. Figure 5.2 shows an overview of anthocyanins (ANs) absorption and excretion.

Previous studies indicate that average quantity of anthocyanins in blood plasma vary from 1.4 and 592 nmol/l followed by post-consumption of dose from 68 and 1300 mg between 0.5–4 h. Further, the average urinary excretion of anthocyanins varies from 0.03 and 4 % of the total ingested anthocyanins, and maximum excretion occurs between 1–4 h and having a $t_{1/2}$ of 1.5–3 h. Although quercetin are excreted slowly ($t_{1/2}$ 11–28 h), which indicates the probability of their bioaccumulation [29], the excretion of anthocyanins is pretty quick ($t_{1/2}$ 1.5–3 h) and no significant bio-accumulation occurs after a diet containing normal amount of anthocyanins. However the importance of increase of colonic metabolites and aglycone metabolic products should be explored.

Pharmacokinetic investigations indicate the presence of anthocyanins and their glucuronide derivatives in blood up to 5 h after intake while transformation into methyl derivatives increases over time (6–24 h). Another study also confirmed the presence of metabolites with their basic anthocyanin skeleton in the urine for up to 24 h [26] thus biological activity changes with metabolic transformation e.g. glycosylation and acylation patterns decrease the bioavailability of an anthocyanin; however, glycosidases present in the GIT may hydrolyze anthocyanins into anthocyanidins, thereby increasing their biological potential but decreasing their stability. The presence of a glucose moiety compared with a galactose or arabinose on the cyanidin and peonidin anthocyanidins present in cranberry juice seems to make them more bioavailable as a percentage of the delivered dose [30].

The absorption of anthocyanins from food is insufficient and the contents detected in plasma are from nM to low μM [26, 28]. Kay et al. [26], reported the total cumulative concentration 172.96±7.44 μg h/mL of anthocyanins (parent and metabolites)

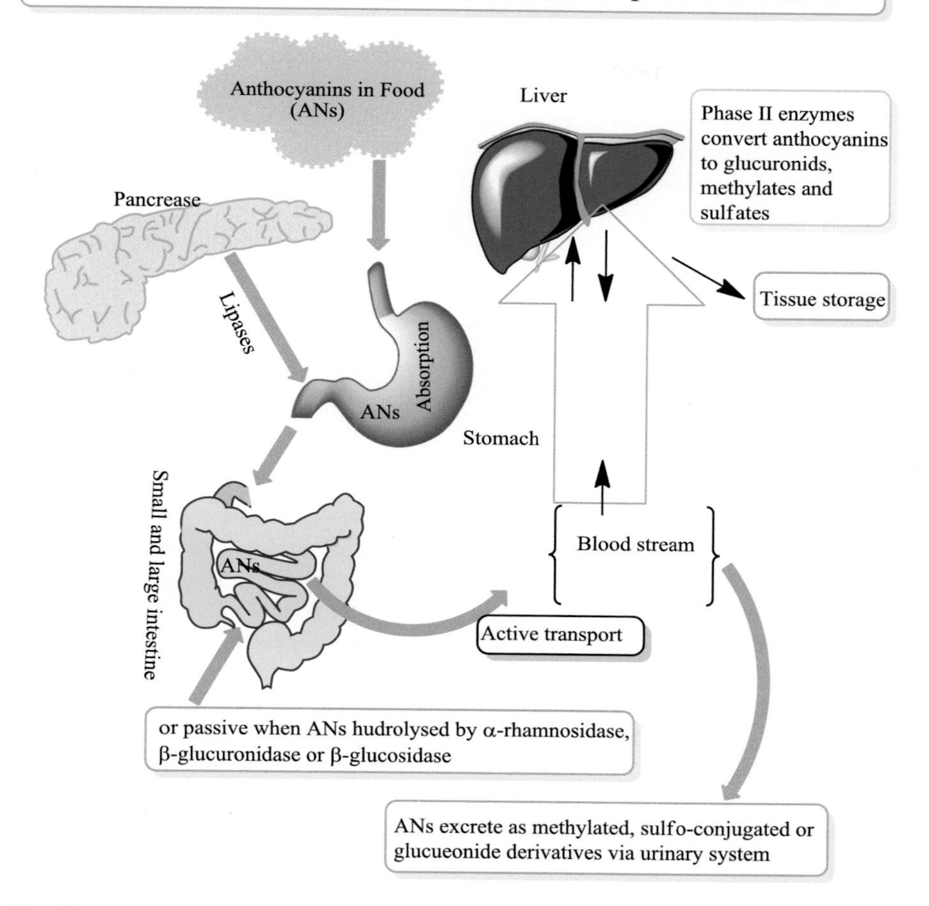

Fig. 5.2 An overview of anthocyanins (ANs) absorption and excretion

detected in the serum over a 7 h sampling, peak plasma level 44.86 ± 2.82 μg/mL was achieved (C_{max}) within 2.8 h (t_{max}), cyanidin 3-glycosides was found the parent of the total anthocyanins in the serum. Further, the total urinary excretion of metabolites and parent compounds over 24 h was 1071.54 ± 375.46 μg, reaching a maximal rate of excretion (R_{max}) of 202.74 ± 85.06 μg/h at 3.72 ± 0.83 h (t_{max}) and having an elimination half-life ($t_{1/2}$) of 4.12 ± 0.4 h. Correspondingly, only 32.5 % of the anthocyanins excreted in the urine were the parent compounds with an average of 67.5 % occurring as conjugated metabolites.

To exert any physiological effect, bioactive molecules should be bioavailable. The percentage of anthocyanins, both native and metabolites, absorbed and evacuated in urine is extremely small in-comparison to consumed amount. The bioavailability of anthocyanins is very low, with <1 % of the ingested amount reaching the plasma, glucuronited and methylated anthocyanin metabolites were found twice than their parent (intact) compounds [14].

The reason for low bioavailability of anthocyanins can be underestimation that may occur because concentration of some metabolites, such as protocatechuic acid, might be below the detection limit of the analytical methods employed and the predominance of the colorless carbinol (75–80 %) and chalcone (15–20 %) forms of anthocyanins present in blood and urine at neutral pHs, it is highly likely that these chemical forms may escape detection, and/or be chemically bound to other components in the plasma or urine, and therefore not included in the analyzed fraction. Labeling anthocyanins for identification of all generated metabolites may overcome these shortcomings [14, 27].

5.4 Carbohydrates Moieties Deconjugation

Glycosylated anthocynains are better hydrophilic as compared to aglycosylated (aglycones) ones and passively diffuse the cell membrane due to their more lipid solubility and small molecular size. During this passage, ingested anthocyanins are deglycosylated. Despite the strong acidic conditions of stomach, there are very little chances of acid hydrolysis (non-enzymatic deglycosylation). Deglycosylation of glycosylated anthocyanins [31] was not observed after pepsin-HCl digestion even at stronger conditions (pH 2.0, 37 °C, 2 h). There are chances of in vivo enzymatic deglycosylation as observed by using animal models although very little evidence is available to support this path. GIT of rats and pigs showed selective degradation of anthocyanin-glucosides in small intestine [2, 20], however further characterization of deglycosylation patterns of anthocyanins under the effect of isolated small intestinal β-glucosidases is needed. Even in rat small intestine in situ perfusion model, the disappearance of cynadin-3-glucoside was significantly higher than other glycosides of cynadin [18]. Limited information available suggests that anthocyanins containing xylose are better retained in the fecal content and feces as opposed to anthocyanins containing galactose and glucose [21]. Detailed studies are required to interpret the destiny of such glycosides.

Various factors influence the bioavailability of anthocyanins, like quantity of gut microflora, food matrix [32], analytical detection problems (e.g. carbinol and chalcone) and differences in xenobiotic metabolism in GIT, liver, and other tissues.

5.5 The Influence of Colonic Microflora

Gut microflora converts anthocyanins into less stable anthocyanidins at neutral pH leading to degradation within 20 min thus Cyanidin-3-rutinoside was first hydrolyzed into cyanidin-3-glucoside and then into cyaniding aglycon, which rapidly degraded into protocatechuic acid (3,4-dihydroxybenzoic acid) [33]. Some anthocyanins metabolites have greater microbial and chemical stability, indicating their potential role in various biological effects and enhancement of antioxidant activity [33]. It is pertinent to mention that degradation of methylated anthocynains by the gut microbiota may yield de-methylated products [27].

Since there are no endogenous esterases in humans to release phenolic acids, therefore the esterase activity of colonic microflora is required for the metabolism of acylated flavonoids [34]. Aura et al. [35] revealed that human fecal flora immediately deconjugates quercitin glucosides, -rutinoside and glucuronide (glc). Simple phenolics like 3,4-dihydroxyphenyl acetic acid and its derivatives are generated during ring fission of deglycosylated quercetin [36]. Aura et al. suggested that gut microflora can convert anthocyanins as revealed by fermentation of cyanidin-3-rutinoside and cyanidin-3-glucoside in the presence of human fecal slurry. Cyanidin-3-glucoside was nearly hydrolyzed entirely after incubation of 2 h, and smaller than one third of the cyanidin-3-rutinoside present. Protocatechuic acid (PC), a ring-fission product of cyanidin-aglycone, was the main metabolite [35]. More than 90 % of the cyanidin-3, 5-glucoside after incubation with human fecal suspension was degenerated after 2 h and fractional hydrolysis yielded cyanidin mono-glucoside as an intermediate of deterioration, which also degraded in the meantime. Resultant production and piling-up of protocatechuic acid was observed again. Assessment of di-acylated anthocyanins from red radish indicated that the acyl group can be broken down by fecal microflora and the released acids were comparatively stable [37]. Deacylated anthocyanins can follow the same pathway of degradation as discussed above.

5.6 Metabolism in Intestinal Mucosa and Tissues

The polyphenolic and cationic attributes of anthocyanins and their metabolites activate a variety of cellular responses. Some metabolites are responsible for bioactivity of flavonoids. For example, methylated cyanidin-3-glucoside is transformed to peonidin-3-glucoside [38]. Similarly, certain metabolic products of anthocyanins may have superior activity as compared to parent molecules.

5.7 Tissue Distribution

The knowledge of tissue distribution of anthocyanins is necessary to evaluate the health-promoting effects of anthocyanins. This information will also help to screen the suitable anthocyanins from various sources to assist development of supplements and functional foods. The protective potential of anthocyanins associated with various tissue-diseases is proved in vitro by using various types of cell lines. However in vivo studies of same are limited. Distribution of anthocyanin in tissues has recently been evaluated in pig and rat models. Male Wistar rats were fed blackberry extract (370 nmol anthocyanin/day) for 15 days and killed at 3 h after the beginning of the last meal. Total anthocyanins averaged 605 nmol g^{-1} in jejunum, 68.6 nmol g^{-1} in stomach, 3.27 nmol g^{-1} in kidney, 0.38 nmol g^{-1} in liver, and 0.25 nmol g^{-1} in brain [28]. In pigs fed diets supplemented with 0, 1, 2, or 4 % w/w blueberries for 4 weeks and fasted for 18–21 h before euthanasia, 1.30 pmol g of anthocyanins were identified in the liver, 1.58 pmol g^{-1} in eyes, 0.878 pmol g^{-1} in cortex, and 0.664 pmol g^{-1} in cerebellum [39]. The results suggested that anthocyanins after crossing the blood-retinal and the blood-brain barrier can protect brain and eye tissues respectively. This hypothesis is also favored by another research in which aged blueberry-fed rats were used [40].

5.8 Excretion

Urine is the major excretory pathway of intact anthocyanins and aglycones [41]. Although most conjugated flavonoid metabolites follow the same route [38], however a small portion of them may re-enter the jejunum via bile, and are absorbed by the colon entering the entero-hepatic circulation again [42], or are defecated with feces. The lung is a chief excretion place for many bioactive constituents including quercetin [43]. However, there is no evidence indicating excretion of anthocyanin by respiration.

Recently biodegradation of anthocyanins into phenolic acids by enteric bacteria has gained much attention (Table 5.1). Various studies have documented the presence of phenolic acids, like protocatechuic acid, gallic acid, vanillic acid, phloroglucinol acid, syringic acid and phloroglucinol aldehyde anthocyanin metabolites [45–47]. These phenolic acids are produced during anthocyanin metabolism either by enteric bacteria, or by chemical conversion, and have also been detected in humans [48].

Anthocyanidin aglycones are degraded much faster as compared to anthocyanins monoglycosides [33]. Similarly anthocyanins degradation by intestinal microflora is also faster [44]. Unabsorbed and non-degraded anthocyanins are excreted in their intact forms. Unchanged anthocyanins are detected in fecal samples of humans 24 h after ingestion of blood orange juice [49], and in fecal samples of rats fed by grapes, bilberries and chokeberries [21].

Table 5.1 Degradation products of anthocyanidin aglycons [44]

Anthocyanidin aglycon		Corresponding phenolic acid
Cyanidin	→	Protocatechuic acid
Delphinidin	→	Gallic acid
Pelargonidin	→	4-Hydroxybenzoic acid
Malvidin	→	Syringic acid
Peonidin	→	Vanilic acid
Petunidin	→	3-O-methylgallic acid

5.9 Conclusions

The beneficial effects of anthocyanins have made the study of their bioavailability, absorption and metabolism an area of considerable interest since humans can't synthesize these vital molecules de novo and obtain them from plant kingdom. Understanding of anthocyanin chemistry, metabolism and their mechanism of action has increased substantially in previous 10 years. Many older questions and controversies have been resolved. Studies have been conducted in last decade to develop a more thorough understanding of factors that influence their uptake, metabolism, and utilization in humans. This information however is only tip of the iceberg. The exact molecular mechanism behind their metabolism remains to be fully established. Although not all animals are good models for metabolic studies of anthocyanins, rats are commonly used. There is need for validation of new animal models especially induced mouse mutants in studying anthocyanin metabolism. The effects of variety, growing practices, season, stage of maturity, storage and processing condition are important for researchers to understand anthocyanin metabolism in epidemiological studies.

References

1. Kühnau, J. (1976). The flavonoids. A class of semi-essential food components: Their role in human nutrition. *World Review of Nutrition and Dietetics, 24*, 117.
2. Wu, X., Pittman, H. E., & Prior, R. L. (2006). Fate of anthocyanins and antioxidant capacity in contents of the gastrointestinal tract of weanling pigs following black raspberry consumption. *Journal of Agricultural and Food Chemistry, 54*(2), 583–589.
3. Mazza, G., & Miniati, E. (1993). *Anthocyanins in fruits, vegetables, and grains*. Boca Raton: CRC.
4. Hertog, M. G., Feskens, E. J., Hollman, P. C., Katan, M. B., & Kromhout, D. (1993). Dietary antioxidant flavonoids and risk of coronary heart disease: The Zutphen Elderly Study. *Lancet, 342*(8878), 1007–1011.
5. Zamora-Ros, R., Andres-Lacueva, C., Lamuela-Raventós, R. M., Berenguer, T., Jakszyn, P., Martínez, C., et al. (2008). Concentrations of resveratrol and derivatives in foods and estimation of dietary intake in a Spanish population: European Prospective Investigation into Cancer and Nutrition (EPIC)-Spain cohort. *British Journal of Nutrition, 100*(01), 188–196.
6. Tsuda, T. (2012). Anthocyanins as functional food factors-chemistry, nutrition and health promotion. *Food Science and Technology Research, 18*(3), 315–324.
7. Prior, R. L., & Gu, L. (2005). Occurrence and biological significance of proanthocyanidins in the American diet. *Phytochemistry, 66*(18), 2264–2280.
8. Hellström, J. K., Torronen, A. R., & Mattila, P. H. (2009). Proanthocyanidins in common food products of plant origin. *Journal of Agricultural and Food Chemistry, 57*(17), 7899–7906.
9. Gu, L., Kelm, M. A., Hammerstone, J. F., Beecher, G., Holden, J., Haytowitz, D., et al. (2004). Concentrations of proanthocyanidins in common foods and estimations of normal consumption. *Journal of Nutrition, 134*(3), 613–617.
10. Santos Buelga, C., & Scalbert, A. (2000). Proanthocyanidins and tannin like compounds – Nature, occurrence, dietary intake and effects on nutrition and health. *Journal of Agricultural and Food Chemistry, 80*(7), 1094–1117.

11. Saura-Calixto, F., Serrano, J., & Goñi, I. (2007). Intake and bioaccessibility of total polyphenols in a whole diet. *Food Chemistry, 101*(2), 492–501.
12. Wang, Y., Chung, S. J., Song, W. O., & Chun, O. K. (2011). Estimation of daily proanthocyanidin intake and major food sources in the US diet. *Journal of Nutrition, 141*(3), 447–452.
13. Cassidy, A., Mukamal, K. J., Liu, L., Franz, M., Eliassen, A. H., & Rimm, E. B. (2013). High anthocyanin intake is associated with a reduced risk of myocardial infarction in young and middle-aged women. *Circulation, 127*(2), 188–196.
14. Mazza, G. (2007). Anthocyanins and heart health. *Annali dell'Istituto Superiore di Sanità, 43*(4), 369.
15. Passamonti, S., Vrhovsek, U., Vanzo, A., & Mattivi, F. (2003). The stomach as a site for anthocyanins absorption from food. *FEBS Letters, 544*(1), 210–213.
16. Talavéra, S., Felgines, C., Texier, O., Besson, C., Lamaison, J. L., & Rémésy, C. (2003). Anthocyanins are efficiently absorbed from the stomach in anesthetized rats. *Journal of Nutrition, 133*(12), 4178–4182.
17. Passamonti, S., Vrhovsek, U., & Mattivi, F. (2002). The interaction of anthocyanins with bilitranslocase. *Biochemical and Biophysical Research Communications, 296*(3), 631–636.
18. Talavéra, S., Felgines, C., Texier, O., Besson, C., Manach, C., Lamaison, J. L., et al. (2004). Anthocyanins are efficiently absorbed from the small intestine in rats. *Journal of Nutrition, 134*(9), 2275–2279.
19. Prior, R. (2004). Absorption and metabolism of anthocyanins: Potential health effects. In M. S. Meskin, W. R. Bidlack, A. J. Davies, D. S. Lewis, & R. K. Randolph (Eds.), *Phytochemicals: Mechanisms of action* (pp. 1–19). Boca Raton: CRC.
20. He, J., Wallace, T. C., Keatley, K. E., Failla, M. L., & Giusti, M. M. (2009). Stability of black raspberry anthocyanins in the digestive tract lumen and transport efficiency into gastric and small intestinal tissues in the rat. *Journal of Agricultural and Food Chemistry, 57*(8), 3141–3148.
21. He, J., Magnuson, B. A., & Giusti, M. M. (2005). Analysis of anthocyanins in rat intestinal contents impact of anthocyanin chemical structure on fecal excretion. *Journal of Agricultural and Food Chemistry, 53*(8), 2859–2866.
22. He, J., Magnuson, B. A., Lala, G., Tian, Q., Schwartz, S. J., & Giusti, M. M. (2006). Intact anthocyanins and metabolites in rat urine and plasma after 3 months of anthocyanin supplementation. *Nutrition and Cancer, 54*(1), 3–12.
23. Felgines, C., Talavéra, S., Gonthier, M. P., Texier, O., Scalbert, A., Lamaison, J. L., et al. (2003). Strawberry anthocyanins are recovered in urine as glucuro-and sulfoconjugates in humans. *Journal of Nutrition, 133*(5), 1296–1301.
24. Kroon, P. A., Clifford, M. N., Crozier, A., Day, A. J., Donovan, J. L., Manach, C., et al. (2004). How should we assess the effects of exposure to dietary polyphenols in vitro? *American Journal of Clinical Nutrition, 80*(1), 15–21.
25. Gee, J. M., DuPont, M. S., Day, A. J., Plumb, G. W., Williamson, G., & Johnson, I. T. (2000). Intestinal transport of quercetin glycosides in rats involves both deglycosylation and interaction with the hexose transport pathway. *Journal of Nutrition, 130*(11), 2765–2771.
26. Kay, C. D., Mazza, G. J., & Holub, B. J. (2005). Anthocyanins exist in the circulation primarily as metabolites in adult men. *Journal of Nutrition, 135*(11), 2582–2588.
27. Wallace, T. C. (2011). Anthocyanins in cardiovascular disease. *Advances in Nutrition, 2*(1), 1–7.
28. Talavéra, S., Felgines, C., Texier, O., Besson, C., Gil-Izquierdo, A., Lamaison, J. L., et al. (2005). Anthocyanin metabolism in rats and their distribution to digestive area, kidney, and brain. *Journal of Agricultural and Food Chemistry, 53*(10), 3902–3908.
29. Manach, C., Williamson, G., Morand, C., Scalbert, A., & Rémésy, C. (2005). Bioavailability and bioefficacy of polyphenols in humans. I. Review of 97 bioavailability studies. *American Journal of Clinical Nutrition, 81*(1), 230S–242S.
30. Milbury, P. E., Vita, J. A., & Blumberg, J. B. (2010). Anthocyanins are bioavailable in humans following an acute dose of cranberry juice. *Journal of Nutrition, 140*(6), 1099–1104.

31. Pérez-Vicente, A., Gil-Izquierdo, A., & García-Viguera, C. (2002). In vitro gastrointestinal digestion study of pomegranate juice phenolic compounds, anthocyanins, and vitamin C. *Journal of Agricultural and Food Chemistry, 50*(8), 2308–2312.
32. Ohnishi, R., Ito, H., Kasajima, N., Kaneda, M., Kariyama, R., Kumon, H., et al. (2006). Urinary excretion of anthocyanins in humans after cranberry juice ingestion. *Bioscience Biotechnology and Biochemistry, 70*(7), 1681–1687.
33. Keppler, K., & Humpf, H.-U. (2005). Metabolism of anthocyanins and their phenolic degradation products by the intestinal microflora. *Bioorganic and Medicinal Chemistry, 13*(17), 5195–5205.
34. Plumb, G. W., Price, K. R., & Williamson, G. (1999). Antioxidant properties of flavonol glycosides from green beans. *Redox Report, 4*(3), 123–127.
35. Aura, A. M., Martin-Lopez, P., O'Leary, K. A., Williamson, G., Oksman-Caldentey, K. M., Poutanen, K., et al. (2005). In vitro metabolism of anthocyanins by human gut microflora. *European Journal of Nutrition, 44*(3), 133–142.
36. Schneider, H., & Blaut, M. (2000). Anaerobic degradation of flavonoids by *Eubacterium ramulus*. *Archives of Microbiology, 173*(1), 71–75.
37. Fleschhut, J., Kratzer, F., Rechkemmer, G., & Kulling, S. E. (2006). Stability and biotransformation of various dietary anthocyanins in vitro. *European Journal of Nutrition, 45*(1), 7–18.
38. Wu, X., Cao, G., & Prior, R. L. (2002). Absorption and metabolism of anthocyanins in elderly women after consumption of elderberry or blueberry. *Journal of Nutrition, 132*(7), 1865–1871.
39. Kalt, W., Blumberg, J. B., McDonald, J. E., Vinqvist-Tymchuk, M. R., Fillmore, S. A., Graf, B. A., et al. (2008). Identification of anthocyanins in the liver, eye, and brain of blueberry-fed pigs. *Journal of Agricultural and Food Chemistry, 56*(3), 705–712.
40. Andres-Lacueva, C., Shukitt-Hale, B., Galli, R. L., Jauregui, O., Lamuela-Raventos, R. M., & Joseph, J. A. (2005). Anthocyanins in aged blueberry-fed rats are found centrally and may enhance memory. *Nutritional Neuroscience, 8*(2), 111–120.
41. McGhie, T. K., Ainge, G. D., Barnett, L. E., Cooney, J. M., & Jensen, D. J. (2003). Anthocyanin glycosides from berry fruit are absorbed and excreted unmetabolized by both humans and rats. *Journal of Agricultural and Food Chemistry, 51*(16), 4539–4548.
42. Ichiyanagi, T., Shida, Y., Rahman, M. M., Hatano, Y., & Konishi, T. (2006). Bioavailability and tissue distribution of anthocyanins in bilberry (*Vaccinium myrtillus* L.) extract in rats. *Journal of Agricultural and Food Chemistry, 54*(18), 6578–6587.
43. Walle, T. (2004). Absorption and metabolism of flavonoids. *Free Radical Biology and Medicine, 36*(7), 829–837.
44. Forester, S. C., & Waterhouse, A. L. (2008). Identification of Cabernet Sauvignon anthocyanin gut microflora metabolites. *Journal of Agricultural and Food Chemistry, 56*(19), 9299–9304.
45. Forester, S. C., & Waterhouse, A. L. (2009). Metabolites are key to understanding health effects of wine polyphenolics. *Journal of Nutrition, 139*(9), 1824S–1831S.
46. Forester, S. C., & Waterhouse, A. L. (2010). Gut metabolites of anthocyanins, gallic acid, 3-O-methylgallic acid, and 2, 4, 6-trihydroxybenzaldehyde, inhibit cell proliferation of Caco-2 cells. *Journal of Agricultural and Food Chemistry, 58*(9), 5320–5327.
47. Ávila, M., María, H., Concepción, S. M., Carmen, P., Teresa, R., & Pascual-Teresa, S. D. (2009). Bioconversion of anthocyanin glycosides by Bifidobacteria and *Lactobacillus*. *Food Research International, 42*(10), 1453–1461.
48. Azzini, E., Vitaglione, P., Intorre, F., Napolitano, A., Durazzo, A., Foddai, M. S., et al. (2010). Bioavailability of strawberry antioxidants in human subjects. *British Journal of Nutrition, 104*(08), 1165–1173.
49. Vitaglione, P., Donnarumma, G., Napolitano, A., Galvano, F., Gallo, A., Scalfi, L., et al. (2007). Protocatechuic acid is the major human metabolite of cyanidin-glucosides. *Journal of Nutrition, 137*(9), 2043–2048.

Chapter 6
Biosynthesis and Stability of Anthocyanins

6.1 Introduction

Anthocyanins biosynthesis is one of the most studied secondary metabolite pathway in plants. It is well-known now that anthocyanins biosynthesis is regulated by several metabolic factors, majority of which are similar across various plant species. Their biosynthetic pathway has been characterized at enzymatic as well as genetic level with gene sequences available for all major biosynthetic steps of anthocyanins. Further various mechanisms that regulate expression of these genes in plant cells have also been studied.

Figure 6.1 summarizes the main steps of anthocyanins biosynthetic pathway. The biosynthesis (phenylpropanoid metabolic pathway) starts from phenylalanine, an amino acid, which yields 4-coumaroyl-CoA. This compound is then pooled with malonyl-CoA to produce chalcones comprising two phenyl rings. The conjugate ring-closure forms the chalcones, the well-known structure of flavonoids, represented by the three-ringed structure (flavanones). Flavanones are subjected to a series of enzymatic modifications that produces dihydroflavonols, anthocyanins but also flavonols, flavan-3-ols, proanthocyanidins (tannins) and other polyphenolics. The dihydro-flavonols are converted to flavonols (myricetin) which further are glycosylated by glycosyl-transferase to the corresponding anthocyanins. Based on the known biosynthetic pathways of anthocyanins, optimal intervention on the involved enzymes of the secondary metabolic pathways may lead to an increase of anthocyanin content of the plant, as shown by research on strawberries [3].

Anthocyanins are produced from precursors by two biosynthetic pathways: shikimate, producing phenylalanine; and that generating malonyl-CoA. These two precursors are interconnected by chalcone synthase by a polyketide folding mechanism, to yield an intermediate chalcone, a substrate for chalcone isomerase producing prototype pigment naringenin which is oxidized by a series of enzymes (flavanone 3-hydroxylase, flavonoid 3′-hydroxylase and flavonoid 3′, 5′-hydroxylase) to leucoanthocyanidins, and finally converted to anthocyanidins by

M. Riaz et al., *Anthocyanins and Human Health*, SpringerBriefs in Food, Health, and Nutrition, DOI 10.1007/978-3-319-26456-1_6

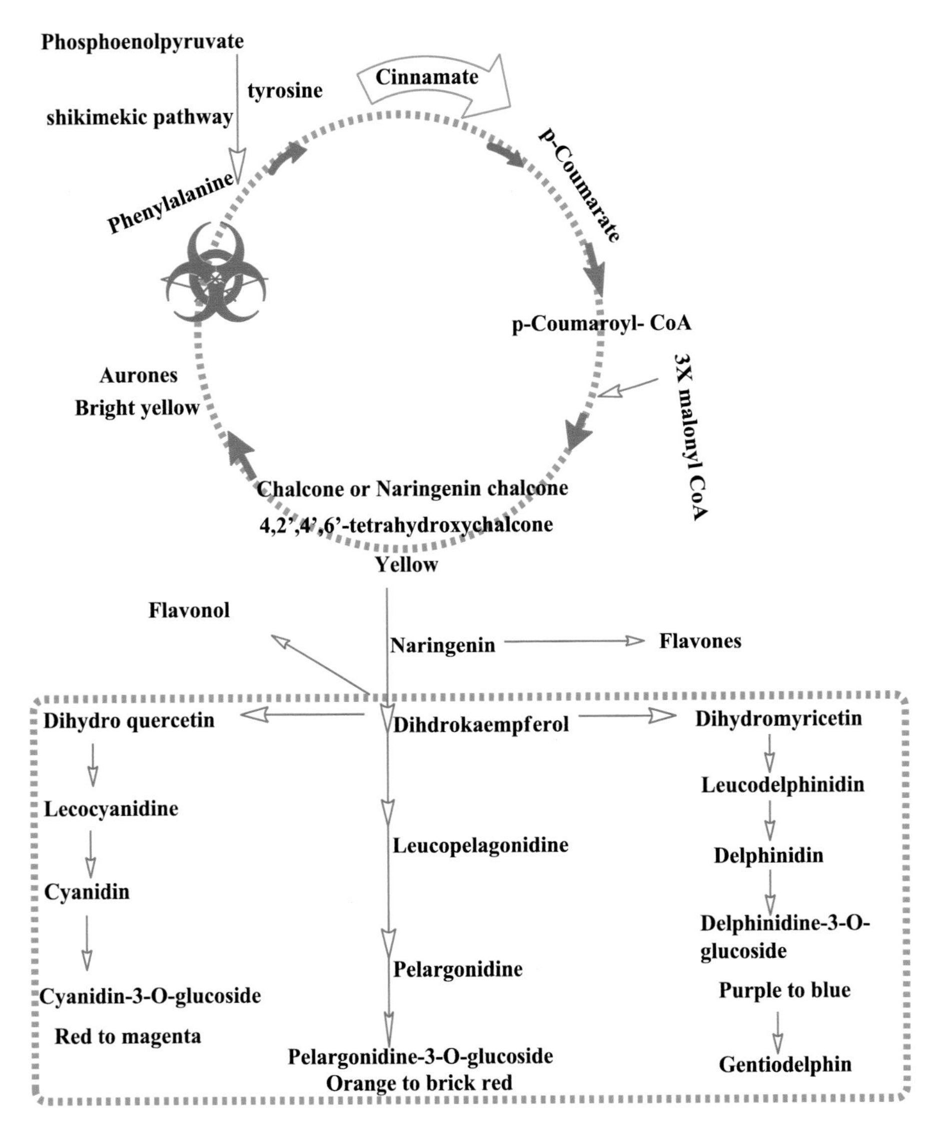

Fig. 6.1 Anthocyanins biosynthetic pathway [1, 2]

leucoanthocyanidin-dioxygenase. These unstable anthocyanidins are joined to various sugars by UDP-glucose/flavonoid 3-*O*-glucosyltransferase and *O*-methyltransferase to yield anthocyanins [4]. They accumulate in vacuoles of cells and tissues of both reproductive and vegetative organs of plants. Out of 17 isolated anthocyanins, majority are found in reproductive organs while 6 aglycones have been identified in vegetative parts of plants. Enzymes involved in biosynthetic pathway of anthocyanins are localized on the cytosolic side of smooth endoplasmic reticulum (SER). From SER, they are transported to cell vacuole by multidrug and toxic compound extrusion (MATE) and ATP-binding cassette (ABC) membrane transported and vesicles.

6.2 Stability of Anthocyanins

Stability of anthocyanins is a major topic of recent research on anthocyanins due to their potential applications, beneficial effects and their use as alternative to artificial colors. More detailed discussion is available about the stability of anthocyanins in their concerned chapter however in product formulation this point may be considered that the stability of anthocyanins increases with increasing the viscosity of their solution and molecular associations within the solution also increase the stability by preventing water based degradations in anthocyanins. Anthocyanins are extremely unstable and very prone to degradation. Their stability depends on oxygen, light, temperature, pH, chemical structure, concentration, solvents, and presence of metal ions, enzymes, proteins and flavonoids [5–8]. Table 6.1 includes composition of various anthocyanins.

Cyanidin and delphinidin are more stable than malvidin, peonidin and petunidin due to the blocking reactive OH group by methylation. Similarly anthocyanins with a 4-substituition are more stable than others. Glycosylation of 3 positions leads to stability while glycosylation of 5 position decreases stability. Acylation of anthocyanins increases their stability by self-association reactions and intra-molecular and intermolecular copigmentation. Therefore acylated anthocyanins can offer the necessary stability for food applications. Acylated anthocyanins can be obtained from black carrots, red cabbage, radish, purple sweet potatoes and red potatoes [5].

Table 6.1 Edible sources of acylated anthocyanins [5]

Source	Type of pigments
Radish (*Raphanus sativus*)	Pelargonidin derivative acylated with one cinnamic acid and an aliphatic acid
Potato (*Solanum tuberosum*)	Pelargonidin derivatives acylated with one cinnamic acid
Black carrot (*Daucus carota* L.)	Cyanidin-3-Rutinoside-Glucoside-Galic acid acylated with one cinnamic acid
Red cabbage (*Brassica oleracea*)	Cyanidin-3-Diglucoside-5-Glucoside acylated with one or two cinnamic acids
Grape (*Vitis labrusca*)	A mixture of five different aglycones, acylated and non-acylated with p-coumaric acid

6.3 Relationships Between Structure and Stability

Anthocyanins are often partially degraded under the joint action of cellular and environmental factors (e.g. light, temperature, pH, metal ions, oxygen, coexisting sugar, etc.). Their molecular structure (e.g., the number and placement of the hydroxyl and methoxyl groups) affects their chemical behavior. The presence of hydroxyl groups on the rings makes the anthocyanins carry a positive charge in acidic solution. Although, the increased hydroxylation could stabilize the anthocyanidin, but excessive hydroxyl groups could also decrease the stability of the pigment molecule, which makes it less stable than those with more methoxyl groups. In the glycosides of anthocyanidin, the types of glycosyl units and acyl groups attached to the aglycone have a significant effect on the structural stability together with their bonding site and number. Acylated anthocyanins are stable due to piling of acyl groups with the pyrylium ring of the flavylium cation, thus protecting the chromophores from nucleophile attack of water leading to synthesis of chalcone or pseudo-base [5]. This idea was also supported by Amiot and his coworkers [9], this phenomenon is termed as intramolecular copigmentation. The color stability is attributed in unacylated anthocyanins to intermolecular copigmentation mechanism (including π–π overlap, dipole–dipole and hydrogen bonding interactions) [7, 10]. Table 6.2 presents a comparative stability of various anthocyanins.

Table 6.2 Comparative stability of anthocyanins [10]

More Stable	Stable	Reference
Pyranoanthocyanins	Anthocyanins > Anthocyanidins	[5, 11]
Disaccharide anthocyanins	Monosaccharidic anthocyanins	[5, 11]
Acylatedanthocyanins	unacylated derivatives anthocyanins	[5, 11]
Polyacylatedanthocyanins	Monoacylated anthocyanins	[5, 11]
Anthocyaninsacylated with aromatic acids (e.g. p-coumaric, caffeic, ferulic acids)	Acylated with aliphatic acids (e.g. acetic, malonic or oxalic acids)	[5, 11]
Acylatedanthocyanins with caffeic acid	Acylated anthocyanins with p-coumaric acid	[5, 11]
aromatic acids substitution in ring B of flavyliumcation	aromatic acids substitution in ring A of flavylium cation	[12]
Color of anthocyanins with petunidin or malvidinaglycones	Color of anthocyanins with pelargonidin, cyanidin, or delphinidinaglycones	[13]
Anthocyanins containing galactose	Anthocyanins containing arabinose	[13]
Anthocyanins at low pH in acidic media	Anthocyaninswith high pH in alkaline solution	[6]

6.4 Factor Affecting Stability of Anthocyanins

6.4.1 pH

Anthocyanins are susceptible toward change in pH [14] in media thus different chemical forms and colors [15]. In acidic aqueous solutions, anthocyanins occur in 4 equilibrium species: the quinonoidal base (QB), the flavylium cation (FC), the carbinol or pseudo-base (PB) and the chalcone (CH) [16, 17]. Table 6.3 indicates the effect of pH on anthocyanins.

Anthocyanidins are less stable at neutral pH while Pg is the most stable anthocyanidin among all, in anthocyanins, monoglycosides and diglycosides derivatives are more stable at pH 7 [18]. It is concluded that color variation with pH is more important in the alkaline region due to their instability [7].

Ozela et al. (2007) evaluated the stability of anthocyanin in the extract of spinach vine fruit (*Basella rubra* L.) in relation to degradative factors such as light, temperature and pH acting alone or in combination. Spinach vine extract is more stable at pH=5.0–6.0 than at pH 4.0, in the absence as well as presence of light. This feature is different from other anthocyanins. This property indicates its potential use as a natural food color [19].

6.4.2 Co-pigmentation Effect

The anthocyanins color and intensity effects are the result of forming molecular or complex associations is a co-pigmentation phenomenon for anthocyanins [20]. Some investigators propose it as stabilization mechanism for anthocyanins color in plants [21]. The electron deficient flavylium ion associate with p-electrons rich copigment thus gives stability against nucleophilic attack of water on 2 position of flavylium [22] and for other species such as $SO_2$and peroxides on 4 position [23]. Figure 6.2 indicates anthocyanin interactions.

The co-pigmentation effect is pH dependent, because at low pH, anthocyanins occur in flavylium form, and at high pH values, they exist in colorless carbinol pseudobase form. The example of charge transfer complex or p–p complex is the anthocyanins and tannins complexation in wines, producing pigmented tannins [24]. Similar reactions occur between flavan-3-ols and tannins [7, 25].

Metals and flavylium salts chelates were considered for variety of colors in flowers [26]. Anthocyanin-metal complexations constitutes possible substitute for color stabilization especially if the metals have no toxic effects. The complex formation occurs with o-di-hydroxyl groups in the B ring (Cy, Dp, Pt) of anthocyanins and anthocyanidins [20]. Some authors suggest that the blue color in plants is due to complexation between anthocyanins and certain metals like Al, Fe, Cu and Sn [27] or Mg and Mo [28]. Anthocyanin-molybdenum complexation causes the

Table 6.3 Effect of pH on anthocyanins [7]

pH of the medium	Predomanint chemical form	Structure
1	Flavylium cation (red color)	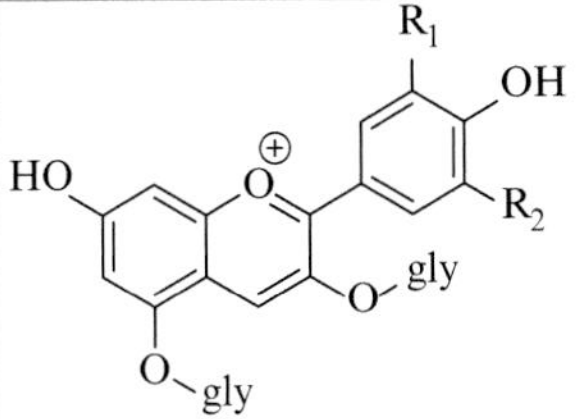 HO, O⊕, O–gly, O–gly, R1, OH, R2
Between 2 and 4	Quinoidal blue species	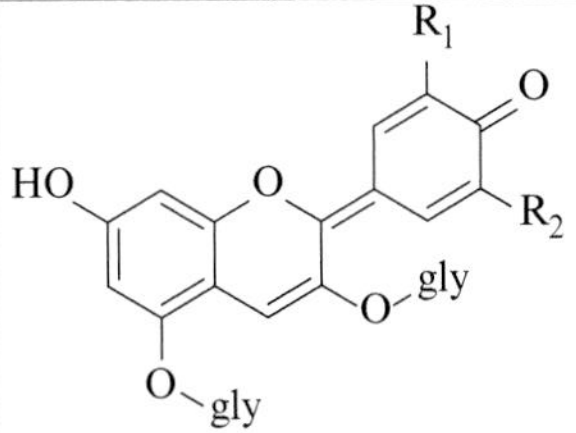 HO, O, O–gly, O–gly, R1, O, R2
Between 5 and 6	Carbinolpseudobase and a chalcone (colorless)	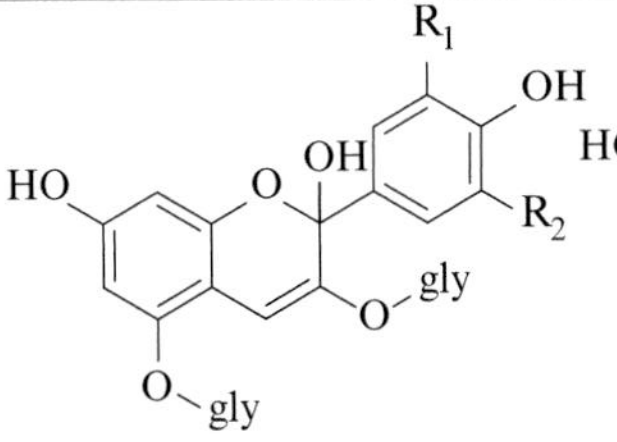 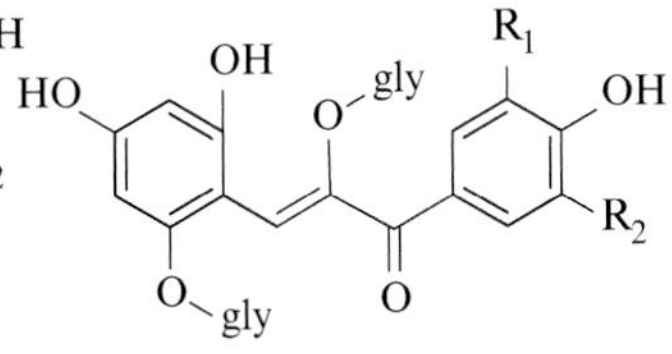 HO, O, OH, O–gly, O–gly, R1, OH, R2 Psedobase carbinol HO, OH, O–gly, O–gly, O, R1, OH, R2 Chalcone
>7	Corresponding degraded products	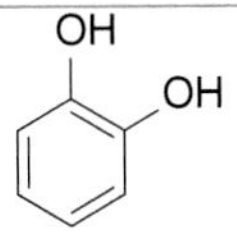 OH, OH Pyrocatechol

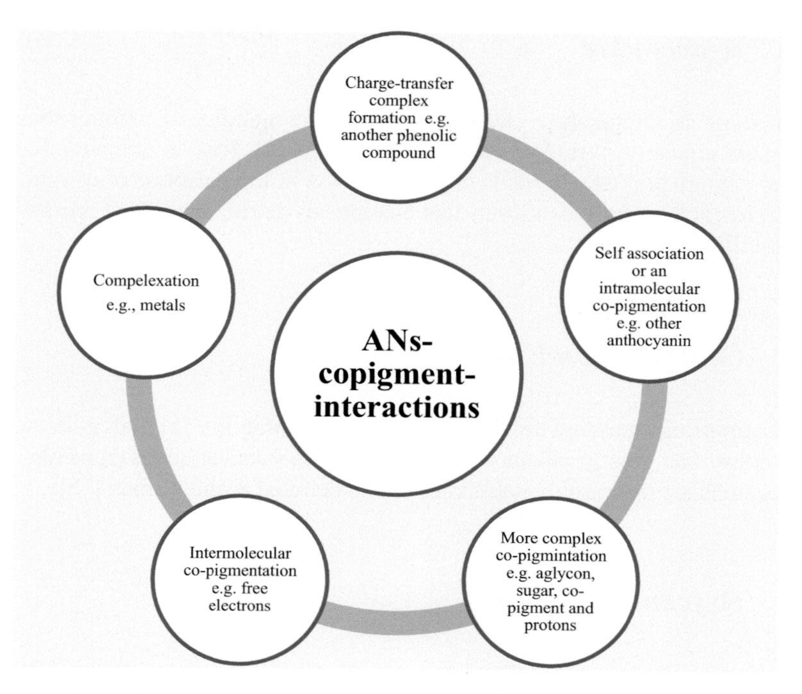

Fig. 6.2 Anthocyanins (ANs) interaction [7]

stabilization of blue color in Hindu cabbage tissue [28]. Yoshida et al. (2006) also confirmed the role of metal complex in plants by complexation between anthocyanins and proved that Mg (II) or Fe (III) at pH 5 are important for the development of blue color in plants [7, 29].

6.4.3 Solvent Effects

Studies on synthetic flavylium salts in different solvents (acetonitrile, dioxane, ethanol, propylene-glycol, water and 2-butanone) have shown that variations in color depend on the flavylium salts concentration and solvent used [30]. In protic solvents, the flavylium salts display red color due to monomer, while in aprotic solvents the solutions are yellow due to dimer while increasing concentration of flavylium salts, the red color is preferred. Another study suggests that increasing H_2O concentration in mixture of water:acetonitrile, the monomer is changed to a green color dimer (monomer with charge-transfer property). Water therefore plays an important role in dimerization of flavylium salts [7].

6.4.4 Temperature

The increase in temperature cause the decrease in stability of anthocyanins and other plant pigments during processing and storage [8]. Rise in temperature cause thermal degradation which yields brown products in the presence of oxygen [31]. Various researchers have confirmed that heating has detrimental effect on anthocyanins stability [32–34].

6.4.5 Concentration Effects

Anthocyanin concentration has direct effect on color stability [5] such as strawberry syrup color was greatly enhanced by mounting the anthocyanins contents. Total anthocyanins are more color stabilizing than individual anthocyanins [35].

6.4.6 Oxygen

The presence of oxygen amplified the degradation process, and the condition is worsening if heating is there. Oxygen and increased temperature were the most harmful combinations ever found in color degradation of various berry juices and isolated anthocyanins [36] demonstrated that oxygen-free environment remained best condition for the antioxidant activity and anthocyanins concentrations of strawberries (fresh-cut) during cold storage. Some investigators have reported that if the atmosphere is enriched with 60–100 % oxygen results increase in phenolic and anthocyanin content during the onset (0–7 days) of cold storage [37]. Yellow colored 3-deoxyanthocyanins (due to dehydroxylation of the carbon at C-3) are more stable than the corresponding red colored 3-hydroxy anthocyanins.

6.4.7 Light

Light is crucial for the biosynthesis of anthocyanins, but it also speeds up their degradation. Dark has safeguarding effect on anthocyanins in comparison to light [38]. Amr and Al-Tamimi [39] in 2007 studied the effect of light on anthocyanins stability; their work verified that light had a very detrimental effect on anthocyanins stability during storage particularly in the presence of sugar. Light exposure cause degradation of anthocyanins in incoherent manner which is source (plant) and anthocyanins dependent degradation [40, 41]. Another study under neon light showed that the phenolic and anthocyanin content was decreased during storage [42].

Ozela et al. (2007) [19] evaluated that various factors like pH, temperature and light affect the anthocyanins stability of the spinach vine fruit (*Basella rubra* L.) extract. Results showed that, independent of pH, the extract exposed to light have anthocyanin degradation kinetics with the average half-life is smaller compared to an average half-life in dark storage. They found that in the presence of light, increased temperature leads to increased anthocyanin degradation.

6.4.8 Enzymes

Glycosidases are the most common anthocyanins degrading enzymes that break the glucosidic linkage and results unstable anthocyanidin [43]. Other anthocyanins degrading enzymes like peroxidases (phenol oxidases) and phenolases (polyphenol oxidases), are present in berries and fruits [44]. Inactivation of enzymes increases anthocyanin stability [45]. Fang et al. (2007) [46] reported that bayberry phenolase alone cannot react with directly with cyaniding-3-glucoside but Mathew and Parpia [47] and Kader et al. (1999) [44], demonstrated that polyphenol oxidase can react directly with anthocyanins although they acted as fairly poor substrates.

6.4.9 Ascorbic Acid

Anthocyanins are degraded in the presence ascorbic acid by direct condensation mechanism [48] and by free radical mechanism the pyrilium ring undergo oxidative cleavage resulting in loss of color of anthocyanins. Ascorbic acid and its degradation products increase the degradation arte of anthocyanins [49].

6.4.10 Sugars

The results of adding sugar to anthocyanin stability depends mainly on the the number and nature of sugars attached to the flavylium cation and number and nature of acids linked to the glycosylic moiety [50]. Sugars and their degradation products decrease anthocyanin stability. The addition of sucrose to mulberry extract increases the stability of anthocyanins by increase in polymeric and co-pigmented anthocyanins during heating before heating the effect was hyperchromic [51]. Similarly anthocyanins content of refrigerated berries extracts were more stabilized by 20 % sugar addition however by increasing the concentration the protective action was decreased [52].

6.4.11 *Sulfites*

Sulphates and sulfites in foods items form colorless sulfur derivatives with anthocyanins which may be reversed by heating or acidification if the concentration of sulfites are not too high (>10 g/kg) [53].

6.5 Stability of Anthocyanins in Food Products

Due to plentiful applications, valuable effects and their use as alternative to artificial colorants in foods; anthocyanins are the main focus of recent investigations. Food processing involves extreme temperature up to 150 °C that may affect the stability of anthocyanins during shelf life. Apart from temperature, other factors like pH, storage temperature, chemical structure, concentration of anthocyanins present, light, oxygen, the presence of enzymes, proteins and metallic ions also affect the shelf lives of anthocyanins enriched products [54]. Generally high temperature may affect concentration of anthocyanins in corresponding food items and products, elevated temperature results in formation of polyphenolic degradation products that may have low antioxidant potentials. Studies are required to confirm whether these thermal induced degradation products have decreased or increased antioxidant capacity. Since the degradation mechanism of anthocyanins is rather complex and perplexing, it is possible that thermal processing could induce some un/expected and un/desired chemical reactions which in directly influence food quality so further investigations are required in this area to predict optimum processing conditions and to opt advance processing rather conventional system for respective anthocyanins products to ensure maximum anthocyanins concentration, and longer shelf lives products to the consumers [54].

6.6 Anthocyanins Degradation in Plants

Anthocyanins degradation is controlled and induced when beneficial to the plant. Anthocyanins degradation is dependent on environmental conditions such as light quality, light intensity and growth temperature. Several enzymes have also been isolated that degrade anthocyanins in postharvest fruit that may be candidates for in vivo degradation. Peroxidase, α-glucosidase and polyphenol oxidase are the main enzyme groups that regulate anthocyanins degradation in fruit juices and extracts [55]. All research studies whether carried out in model systems or in extracts show that anthocyanin degradation follows first order kinetics.

Anthocyanins usually accumulate in young leaves and degrade as the leaves become mature [56]. Anthocyanins protect the leaves especially their photosynthetic apparatus from harmful UV light as well as from photo-inhibitory high intensities of visible light [57]. They also act as antioxidants thus protecting the cells from oxidative damage [58]. The decrease in anthocyanins content may be due to increase chlorophyll accumulation as leaf expand and grow, and no longer is photoprotection required [55].

Anthocyanins degradation is also observed in developing fruit e.g. *Capsicum* spp. and Sicilian sweet orange varieties. Anthocyanins accumulation in fruit protects the photosynthetic apparatus in the developing fruit [55].

Flowers often change color during development, acting as a signal for pollinators. In most cases, the change in color is due to induction of anthocyanins synthesis, but in others, such as *Brunfelsia calycina*, anthocyanin is degraded, resulting in a change of flower color from dark purple to white after synthesis [59].

6.7 Drawback and Derivatives of Anthocyanins

As previously discussed, anthocyanins are unstable and their colors varies depending upon light, temperature, pH and presence of metals in media [60]. There are some limitations like decreased stability to processing, storage and formulation conditions which can impart un-wanted flavor or odor characteristics that have limited their use as natural colorants in food systems. The area required attention to developed processes and mechanisms to reduce the incompatibility of anthocyanins with pH, temperature and other environmental changes. For example there may be few coloring principles in red cabbage and radish extracts so techniques and procedures must be adopted to isolated these few necessary components as coloring agents [5].

6.8 Anthocyanins Stabilization Mechanisms

As mentioned before, naturally anthocyanins are unstable, however they can be stabilized i.e. by applying various mechanisms like acetylations, association (co-pigmentation) and microencapsulation when a "co-pigment" is added. The stabilization of grape skin anthocyanins by co-pigmentation with enzymatically modified isoquercitrin (EMIQ) was increased significantly [61].

Encapsulation by spray-drying is a cost-effective way for protection of anthocyanins by coating the ingredients [62]. Thus bioavalability and the shelf lives of anthocyanins may be improved by providing protection against oxygen, water and light [53]. The encapsulation from diverse sources is given in Table 6.4 and the details of steps during process and their advantages are also given in Table 6.5.

Table 6.4 Anthocyanins encapsulation from various sources [53]

Anthocyanin source	Encapsulation technique	Coating material	Stabilization improvement
Concord grape	Spray drying	Maltodextrin	Not measured
Cranberry pomace	Spray drying	Maltodextrin	Not measured
Roselle calyces	Spray drying	Maltodextrin	Not measured
Roselle calyces	Inclusion complexation	β-cyclodextrin	The addition of β-cyclodextrin does not ↑ anthocyanin stabilization
Apple pomace	Spray drying	Maltodextrin	Anthocyanins have↑ stability The ↑ in the shelf life is due to ↓ water activity
Roselle calyces	Liophilization	Pullulan	Encapsulated anthocyanins have slightly ↑ stability as compared to the free anthocyanin extract
Black currant	Thermal gelation	Glucan gel	Encapsulation of anthocyanins followed by appropriate processing ↑ the stability of anthocyanins
Black carrot	Spray drying	Maltodextrin	Not measured
'Isabel' grape bagasse	Spray drying	Maltodextrin and Arabic gum	Not measured
Blackberry	Thermal gelation	Sodium alginate and pectin	Not measured
Blackberry	Ionic gelation	Curdlan	Not measured
Jabuticaba fruit skin	Ionic gelation	Sodium alginate	Not measured

Table 6.5 Some common microencapsulation processes for anthocyanins characteristics [63]

Technology	Particle size (μm)	Advantages	Disadvantages
Spray-drying	10–400	Low process cost, wide choice of wall material, good encapsulation efficiency, good stability of the finished product and potential of industrial production continuously	Can degrade at highly temperature-sensitive compounds, control of the particle size is difficult and yields for small batches are moderate
Freeze drying	20–5000	Best for thermo-sensitive substances in aqueous solutions	More processing time, costly process, costly transport and storage of the capsules
Fluid bed coating	20–200	Low cost process and allows specific capsule size distribution and low porosities into the product	Degradation of temperature-sensitive molecules

(continued)

Table 6.5 (continued)

Technology	Particle size (μm)	Advantages	Disadvantages
Emulsification	0.2–5000	Polar, non-polar (apolar), and amphiphilic can be incorporated	Low number of emulsifiers that can be utilized and hard to control formation of capsules
Melt extrusion	300–5000	The material is completely shielded by the wall material, remaining core is washed externally, low temperature entrapping assay	The capsule should be essentially separated from liquid bath and dried Hard to obtain capsules in highly viscous carrier materials melts
Rapid expansion of supercritical fluid (RESS)	10–400	Easy removal of solvent and non-toxic method, works at ↓temperature and in in-active atmosphere thereby avoiding product degradation	Both the core and the wall material must be very soluble in supercritical fluids, low or no solubility of high molecular weight, polar compounds CO_2, poor control over the precipitated crystal morphology and size distribution
Ionic Gelation		Organic solvents and extreme conditions of temperature and pH are avoided	Mostly used on lab-scale The capsules are highly porous which leads to intensive burst
Thermal Gelation		The same of ionic gelation	Same as mentioned above

6.9 Conclusions

A lot of data is available on molecular biology of anthocyanin biosynthesis. The genes regulating the various steps and the influence of phytohormones on this genetic regulation are further clarified now. Their intra-cellular transportation and accumulation are also better understood. Although substantial progress has been made in our understanding of anthocyanin biosynthesis in plants, many key issues are yet to be addressed in detail. For example the interaction between anthocyanin biosynthetic pathway and other pathways and the effects of such interaction on plant growth and development are not fully understood. It is hoped that next decade will clarify remaining ambiguities in this regard.

References

1. De Pascual-Teresa, S., & Sanchez-Ballesta, M. T. (2008). Anthocyanins: From plant to health. *Phytochemistry Reviews, 7*(2), 281–299.
2. Teixeira da Silva, J. A., Serena, A., Wei, L., Hao, Y., & Akira, K. (2014). Genetic control of flower development, color and senescence of *Dendrobium orchids. Scientia Horticulturae, 175*, 74–86.

3. Oancea, S., & Oprean, L. (2011). Anthocyanin extracts in the perspective of health benefits and food applications. *Revista de Economia, 218.*
4. Kassim, A., Poette, J., Paterson, A., Zait, D., McCallum, S., Woodhead, M., et al. (2009). Environmental and seasonal influences on red raspberry anthocyanin antioxidant contents and identification of quantitative traits loci (QTL). *Molecular Nutrition and Food Research, 53*(5), 625–634.
5. Giusti, M. M., & Wrolstad, R. E. (2003). Acylated anthocyanins from edible sources and their applications in food systems. *Biochemical Engineering Journal, 14*(3), 217–225.
6. Rein, M. (2005). *Copigmentation reactions and color stability of berry anthocyanins*. Helsinki: University of Helsinki.
7. Castañeda-Ovando, A., Lourdes Pacheco-Hernández, M. D. L., Páez-Hernández, M. E., José, A. R., & Galán-Vidal, C. A. (2009). Chemical studies of anthocyanins: A review. *Food Chemistry, 113*(4), 859–871.
8. Mercadante, A. Z., & Bobbio, F. O. (2008). Anthocyanins in foods: Occurrence and physicochemical properties. In C. Socaciu (Ed.), *Food colorants: Chemical and functional properties* (pp. 241–276). Boca Raton: CRC.
9. Malien-Aubert, C., Dangles, O., & Amiot, M. J. (2001). Color stability of commercial anthocyanin-based extracts in relation to the phenolic composition. Protective effects by intra-and intermolecular copigmentation. *Journal of Agricultural and Food Chemistry, 49*(1), 170–176.
10. Li, H., Zeyuan, D., Honghui, Z., Chanli, H., Ronghua, L., Christopher, Y. J., et al. (2012). Highly pigmented vegetables: Anthocyanin compositions and their role in antioxidant activities. *Food Research International, 46*(1), 250–259.
11. Stintzing, F. C., & Carle, R. (2004). Functional properties of anthocyanins and betalains in plants, food, and in human nutrition. *Trends in Food Science and Technology, 15*(1), 19–38.
12. Yoshida, K., Reiko, O., Kiyoshi, K., & Tadao, K. (2002). Prevention of UV-light induced E, Z-isomerization of caffeoyl residues in the diacylated anthocyanin, gentiodelphin, by intramolecular stacking. *Tetrahedron Letters, 43*(35), 6181–6184.
13. Von Elbe, J., & Schwartz, S. (1996). Colorants. *Food Chemistry, 3*, 651–723.
14. Wesche-Ebeling, P., & Argaiz-Jamet, A. (2002). Stabilization mechanisms for anthocyanin the case for copolymerization reactions. In J. Welti-Chanes, G. V. Barbosa-Cánovas, & J. M. Aguilera (Eds.), *Engineering and food for the 21st century* (pp. 141–150). Boca Raton: CRC.
15. Brouillard, R., & Markakis, P. (1982). *Anthocyanins as food colors*. New York: Academic.
16. da Costa, C. T., Bryant, C. N., Sam, A. M., & Derek, H. (1998). Separation of blackcurrant anthocyanins by capillary zone electrophoresis. *Journal of Chromatography A, 799*(1), 321–327.
17. Kennedy, J. A., & Waterhouse, A. L. (2000). Analysis of pigmented high-molecular-mass grape phenolics using ion-pair, normal-phase high-performance liquid chromatography. *Journal of Chromatography A, 866*(1), 25–34.
18. Fleschhut, J., Kratzer, F., Rechkemmer, G., & Kulling, S. E. (2006). Stability and biotransformation of various dietary anthocyanins in vitro. *European Journal of Nutrition, 45*(1), 7–18.
19. Ferreira Ozela, E., Stringheta, P. C., & Cano Chauca, M. (2007). Stability of anthocyanin in spinach vine (*Basella rubra*) fruits. *Ciencia e Investigación Agraria, 34*(2), 115–120.
20. Boulton, R. (2001). The copigmentation of anthocyanins and its role in the color of red wine: A critical review. *American Journal of Enology and Viticulture, 52*(2), 67–87.
21. Davies, A., & Mazza, G. (1993). Copigmentation of simple and acylated anthocyanins with colorless phenolic compounds. *Journal of Agricultural and Food Chemistry, 41*(5), 716–720.
22. Matsufuji, H., Otsuki, T., Takeda, T., Chino, M., & Takeda, M. (2003). Identification of reaction products of acylated anthocyanins from red radish with peroxyl radicals. *Journal of Agricultural and Food Chemistry, 51*(10), 3157–3161.
23. Garcia-Viguera, C., & Bridle, P. (1999). Influence of structure on color stability of anthocyanins and flavylium salts with ascorbic acid. *Food Chemistry, 64*(1), 21–26.
24. Mirabel, M., Saucier, C., Guerra, C., & Glories, Y. (1999). Copigmentation in model wine solutions: Occurrence and relation to wine aging. *American Journal of Enology and Viticulture, 50*(2), 211–218.

25. Salas, E., Atanasova, V., Poncet-Legrand, C., Meudec, E., Mazauric, J. P., & Cheynier, V. (2004). Demonstration of the occurrence of flavanol-anthocyanin adducts in wine and in model solutions. *Analytica Chimica Acta, 513*(1), 325–332.
26. Clifford, M. N. (2000). Anthocyanins-nature, occurrence and dietary burden. *Journal of the Science of Food and Agriculture, 80*(7), 1063–1072.
27. Starr, M., & Francis, F. (1973). Effect of metallic ions on color and pigment content of cranberry juice cocktail. *Journal of Food Science, 38*(6), 1043–1046.
28. Hale, K. L., McGrath, S. P., Lombi, E., Stack, S. M., Terry, N., Pickering, I. J., et al. (2001). Molybdenum sequestration in brassica species. A role for anthocyanins? *Plant Physiology, 126*(4), 1391–1402.
29. Yoshida, K., Kitahara, S., Ito, D., & Kondo, T. (2006). Ferric ions involved in the flower color development of the Himalayan blue poppy, *Meconopsis grandis. Phytochemistry, 67*(10), 992–998.
30. Ito, F., Nobuaki, T., Akio, K., & Tsuneo, F. (2002). Why do flavylium salts show so various colors in solution?: Effect of concentration and water on the flavylium's color changes. *Journal of Photochemistry and Photobiology A: Chemistry, 150*(1), 153–157.
31. Markakis, P. (2012). *Anthocyanins as food colors*. New York: Elsevier.
32. Jiménez, N., Bohuon, P., Lima, J., Dornier, M., Vaillant, F., & Pérez, A. M. (2010). Kinetics of anthocyanin degradation and browning in reconstituted blackberry juice treated at high temperatures (100–180°C). *Journal of Agricultural and Food Chemistry, 58*(4), 2314–2322.
33. Lin, Y.-C., & Chou, C.-C. (2009). Effect of heat treatment on total phenolic and anthocyanin contents as well as antioxidant activity of the extract from *Aspergillus awamori*-fermented black soybeans, a healthy food ingredient. *International Journal of Food Science and Nutrition, 60*(7), 627–636.
34. Sadilova, E., Florian, C., Stintzing, D. R., & Kammerer, R. C. (2009). Matrix dependent impact of sugar and ascorbic acid addition on color and anthocyanin stability of black carrot, elderberry and strawberry single strength and from concentrate juices upon thermal treatment. *Food Research International, 42*(8), 1023–1033.
35. Skrede, R. E. G., Wrolstad, P. L., & Enersen, G. (1992). Color stability of strawberry and blackcurrant syrups. *Journal of Food Science, 57*(1), 172–177.
36. Odriozola-Serrano, I., Soliva-Fortuny, R., & Martín-Belloso, O. (2010). Changes in bioactive composition of fresh-cut strawberries stored under superatmospheric oxygen, low-oxygen or passive atmospheres. *Journal of Food Composition and Analysis, 23*(1), 37–43.
37. Zheng, Y., Shiow, Y. W., Chien, Y. W., & Wei, Z. (2007). Changes in strawberry phenolics, anthocyanins, and antioxidant capacity in response to high oxygen treatments. *LWT – Food Science and Technology, 40*(1), 49–57.
38. Kearsley, M., & Rodriguez, N. (1981). The stability and use of natural colors in foods: Anthocyanin, β-carotene and riboflavin. *International Journal of Food Science and Technology, 16*(4), 421–431.
39. Amr, A., & Al Tamimi, E. (2007). Stability of the crude extracts of *Ranunculus asiaticus* anthocyanins and their use as food colorants. *International Journal of Food Science and Technology, 42*(8), 985–991.
40. Inami, O., Itaru, T., Hiroe, K., & Nobuji, N. (1996). Stability of anthocyanins of *Sambucus canadensis* and *Sambucus nigra. Journal of Agricultural and Food Chemistry, 44*(10), 3090–3096.
41. Shi, Z., Bassa, I. A., Gabriel, S. L., & Francis, F. J. (1992). Anthocyanin pigments of sweet potatoes–*Ipomoea batatas. Journal of Food Science, 57*(3), 755–757.
42. Maier, T., Matthias, F., Andreas, S., Dietmar, R., & Kammerer, R. C. (2009). Process and storage stability of anthocyanins and non-anthocyanin phenolics in pectin and gelatin gels enriched with grape pomace extracts. *European Food Research and Technology, 229*(6), 949–960.
43. Huang, H. (1956). The kinetics of the decolorization of anthocyanins by fungal "anthocyanase" 1. *Journal of the American Chemical Society, 78*(11), 2390–2393.
44. Kader, F., Irmouli, M., Zitouni, N., Nicolas, J. P., & Metche, M. (1999). Degradation of cyanidin 3-glucoside by caffeic acid o-quinone. Determination of the stoichiometry and characterization of the degradation products. *Journal of Agricultural and Food Chemistry, 47*(11), 4625–4630.

45. Garcia-Palazon, A., Suthanthangjai, W., Kajda, P., & Ioannis, Z. (2004). The effects of high hydrostatic pressure on β-glucosidase, peroxidase and polyphenoloxidase in red raspberry (*Rubus idaeus*) and strawberry (*Fragaria ananassa*). *Food Chemistry, 88*(1), 7–10.
46. Fang, Z., Min, Z., Yunfei, S., & Jingcai, S. (2007). Polyphenol oxidase from bayberry (*Myrica rubra* Sieb. et Zucc.) and its role in anthocyanin degradation. *Food Chemistry, 103*(2), 268–273.
47. Mathew, A., & Parpia, H. (1971). Food browning as a polyphenol reaction. *Advances in Food Research, 19*, 75–145.
48. Poei-Langston, M., & Wrolstad, R. (1981). Color degradation in an ascorbic acid anthocyanin flavanol model system. *Journal of Food Science, 46*(4), 1218–1236.
49. Pacheco-palencia, L. A., Hawken, P., & Talcott, S. T. (2007). Juice matrix composition and ascorbic acid fortification effects on the phytochemical, antioxidant and pigment stability of açai (*Euterpe oleracea* Mart.). *Food Chemistry, 105*(1), 28–35.
50. Rubinskiene, M., Viskelis, P., Jasutiene, I., Viskeliene, R., & Bobinas, C. (2005). Impact of various factors on the composition and stability of black currant anthocyanins. *Food Research International, 38*(8), 867–871.
51. Tsai, P. J., Delva, L., Yu, T. Y., Huang, Y. T., & Dufosse, L. (2005). Effect of sucrose on the anthocyanin and antioxidant capacity of mulberry extract during high temperature heating. *Food Research International, 38*(8), 1059–1065.
52. Nikkhah, E., Khayamy, M., Heidari, R., & Jamee, R. (2007). Effect of sugar treatment on stability of anthocyanin pigments in berries. *Journal of Biological Sciences, 7*(8), 1412–1417.
53. Cavalcanti, R. N., Santos, D. T., & Meireles, M. A. A. (2011). Non-thermal stabilization mechanisms of anthocyanins in model and food systems – An overview. *Food Research International, 44*(2), 499–509.
54. Patras, A., Nigel, P. B., O'Donnell, C., & Tiwari, B. K. (2010). Effect of thermal processing on anthocyanin stability in foods; mechanisms and kinetics of degradation. *Trends in Food Science and Technology, 21*(1), 3–11.
55. Oren-Shamir, M. (2009). Does anthocyanin degradation play a significant role in determining pigment concentration in plants? *Plant Science, 177*(4), 310–316.
56. Chalker Scott, L. (1999). Environmental significance of anthocyanins in plant stress responses. *Photochemistry and Photobiology, 70*(1), 1–9.
57. Steyn, W. J., Wand, S. J. E., Holcroft, D. M., & Jacobs, G. (2002). Anthocyanins in vegetative tissues: A proposed unified function in photoprotection. *New Phytologist, 155*(3), 349–361.
58. Winkel-Shirley, B. (2002). Biosynthesis of flavonoids and effects of stress. *Current Opinion in Plant Biology, 5*(3), 218–223.
59. Vaknin, H., Bar-Akiva, A., Ovadia, R., Nissim-Levi, A., Forer, I., Weiss, D., et al. (2005). Active anthocyanin degradation in Brunfelsia calycina (yesterday–today–tomorrow) flowers. *Planta, 222*(1), 19–26.
60. Da Costa, C. T., Horton, D., & Margolis, S. A. (2000). Analysis of anthocyanins in foods by liquid chromatography, liquid chromatography-mass spectrometry and capillary electrophoresis. *Journal of Chromatography A, 881*(1), 403–410.
61. Valentová, K., Vrba, J., Bancířová, M., Ulrichová, J., & Křen, V. (2014). Isoquercitrin: Pharmacology, toxicology, and metabolism. *Food and Chemical Toxicology, 68*, 267–282.
62. Cai, Y., & Corke, H. (2000). Production and properties of spray dried amaranthus betacyanin pigments. *Journal of Food Science, 65*(7), 1248–1252.
63. Mahdavi, S. A., Seid, M. J., Mohammad, G., & Elham, A. (2014). Spray-drying microencapsulation of anthocyanins by natural biopolymers: A review. *Drying Technology, 32*(5), 509–518.

Chapter 7
The Role of Anthocyanins in Health as Antioxidant, in Bone Health and as Heart Protecting Agents

7.1 Introduction

Anthocyanins are one of the most well-known sub-groups of pigmented flavonoid compounds in the plant kingdom. They can be found in almost all land plants, but found in relatively high levels in 27 plant families. The dietary consumption of anthocyanins is high due to their occurrence in fruits and vegetables. They function as phytoprotective substances, have a role in plant, animal interactions and as such are important in eco-physiology or plant defense mechanisms. Recently, interest in anthocyanins as well as in their biological and therapeutic properties has strongly increased. Nemours in vitro studies, animal models and human clinical trials support their medicinal benefits to human. These studies suggest that anthocyaninshave anti-carcinogenic and anti-inflammatory effects, provide cardiovascular disease prevention, promote obesity and diabetes control benefits, and also improve visual and brain functions. To understand the role of anthocyanins in prevention of diseases, it is essential to have a working-knowledge of anti-oxidants and their regulatory role on the human immune system.

Reactive species is a collective term used to indicate oxygen radicals, nitrogen radicals and their associated nonradicals species that act as oxidizing agents. These species are known as reactive oxygen species (ROS) and reactive nitrogen species (RNS) respectively or commonly as oxidants. These radicals can be produced by a variety of mechanism in human body and their uncontrolled production can be harmful for biomolecules, cells, tissues and body organs known as oxidative damage. The in vivo generation of ROS and RNS elicits the radical scavenging or antioxidant mechanism collectively known as antioxidant defense which minimizes and repairs this oxidative damage. This oxidant-antioxidant balance is required for proper functioning of cells and during disease; this balance is sloped in favor of oxidants thus generating oxidative stress and increasing oxidative damage. The ROS and RNS are involved in pathology of more than 100 various diseases and metabolic disorders. This damage may be countered by oral administration of

M. Riaz et al., *Anthocyanins and Human Health*, SpringerBriefs in Food, Health, and Nutrition, DOI 10.1007/978-3-319-26456-1_7

antioxidants from antioxidant rich sources of fruits, vegetables, nuts, grains and cereals. It has led to new antioxidant based preventive/curative nutrition based therapies. These antioxidants modulate innate as well as adaptive immunity and reverse various age-related immune deficiencies by stimulating and boosting the human immune system. Although more significant effects have been noted in elderly, the antioxidant rich diets have same effects in youngers. It is therefore recommended to have a sufficient intake of antioxidant rich foods regularly to prevent and delay the onset of various age-related degenerative disorders. Some of these food-related antioxidants are ascorbic acid, tocopherols, carotenoids and anthocyanins.

It is well-established now that anthocyanin stimulates immunity against tumor growth. Anthocyanin intervention can be helpful strategy in modulating the immune response. Whether the immune modulatory effects of anthocyanins translate into health benefits should be studied with various biological and animal models. It must be kept in mind that these studies have their drawbacks as well. For example, studies using one pure anthocyanin, the aspect of interaction of various dietary bioactive constituents and complete array of various anthocyanin are missing. Similarly for studies involving plant material or extract, which contain collection of other constituents and so results can't be attributed to anthocyanin only.

Different signaling pathways like nuclear factor κB, Wnt/β-catenin, AMP-activated and mitogen-activated protein kinase as well as certain central cellular processes, such as cell cycle, apoptosis, autophagy and biochemical metabolism are involved in these valuable effects of anthocyanins and can provide potential therapeutic strategies for treatment of broad spectrum of diseases in future [1]. Figure 7.1 gives a general overview of pharmacological effects of anthocyanins on various organs and tissues of human body.

7.2 Presumed Health-Promoting Effects of Anthocyanins

Due to their health-promoting and immunity-boosting properties, anthocyanins are appropriate nutraceuticals and supplementary treatments for various aspects of chronic diseases. Dieto-therapeutic applications of anthocyanin-rich foods optimize health and performance constituting an important largest market of nutraceutical products throughout the world. Therefore, it is strongly recommended to identify these health-promoting components in anthocyanin extracts that led to new opening of opportunities of using these extracts in a variety of food applications.

A lot of research has been performed during last decade to identify non-essential nutrients including anthocyanin responsible for observed health-benefits and to verify various claimed physiological effects on human body. Their beneficial effects include anti-inflammatory and anti-cancer activities and protective effects against various metabolic, degenerative and cardiovascular diseases and vision improvements. There are numerous studies in animal/human cell lines, animal models and

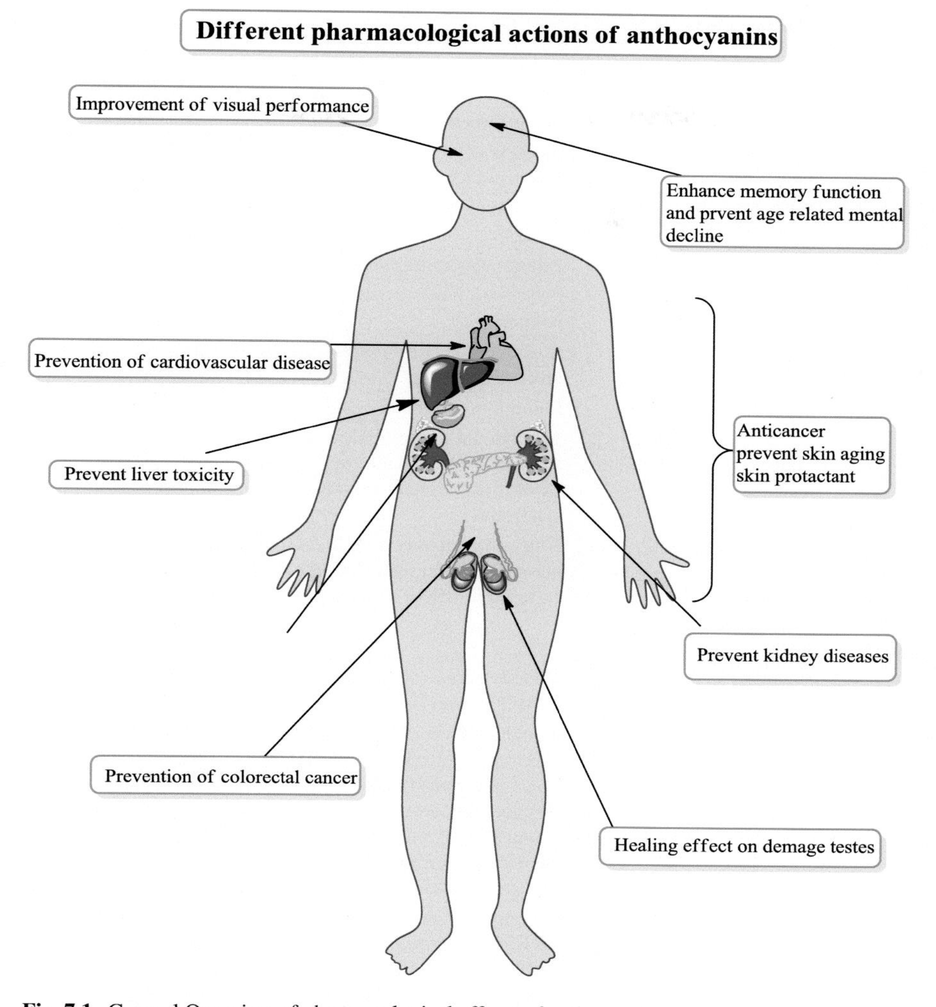

Fig. 7.1 General Overview of pharmacological effects of anthocyanins

human clinical trials that support above mentioned effects of anthocyanins. Their peculiar chemical diversity, suitable molecular weight, specific three dimensional conformation and resulting physical and biochemical effects allow them to interact with various targets in cells producing desirable effects. Data retrieved from various studies support the health-promoting and immunity boosting effects of anthocyanins. Their and their metabolites role in maintaining optimal functions of various physiological and biochemical processes is supported by epidemiological studies, in vitro studies, cell-culture studies and recently in vivo studies. These efforts have led to better understanding of the role of anthocyanin and their derivatives in the

Table 7.1 Healthy effects of uses of anthocyanins

Anthocyanins	Health use	References
Anthocyanins extracts	Sight acuteness	[2]
	Enhancement of antioxidant capacity	[3, 4]
	Cured several blood circulation ailments due to capillary fragility	[5]
	Vaso-protective and anti-inflammatory effects	[6]
	Platelet aggregation inhibition effects	[7]
	Preservation of normal vascular permeability	[5]
	Anti-diabetic, anti-neoplastic and chemo-protective effects	[8, 9]
	Radiation-protective agents	[10]
	Antioxidant properties and notable effects against Cardiovascular hypertension, chronic inflammation, Cancer and metabolic syndrome	[11, 12]
Red orange extract (*Citrus sinensis* varieties: Moro, Tarocco, Sanguinello) (Anthocyanins, flavanones, hydroxycinnamic acids, and ascorbic acid)	Anti-inflammatory activity was assessed in human keratinocytes (lineage NCTC 2544) exposed to IFN-γ and histamine	[13, 14]
Red orange extract at various concentrations	Halted expression of ICAM-1 and secretion of MCP-1 and IL-8	[13, 14]
Grape seed proanthocyanidins	Human keratinocytes irradiated with UVB and treated with GSP's inhibited formation of UVB-induced H_2O_2, protein oxidation, lipid peroxidation, DNA damage and exhaustion of antioxidant components like catalase, superoxide dismutase, glutathione peroxidase and glutathione	[14, 15]
Grape seed proanthocyanidins	Prevented phosphorylation of JNK, p38, ERK1/2and proteins of MAPK family and UVB-induced activation of NF-κB/p65. It suggests that GSP canreduce UV-induced oxidative stress in human skin	[14, 15]

process of various chronic diseases, with special focus to their metabolism and biological actions, dose-effects and organ specific effects. Table 7.1 indicates healthy effects of uses of anthocyanins on various tissues, organs and organ systems of human body.

Interest in anthocyanins has increased after the recognition of their health benefits [16]. Due to their anti-inflammatory and anti-oxidant potential, epidemiologic studies suggest that ingestion of anthocyanins reduces the risk of diabetes, arthritis, cancer and cardiovascular disease [17]. Their effects are due to their effects at various molecular levels by directly scavenging ROS [18] or inducing phase II detoxifying enzymes or to induce phase II detoxifying enzymes [19].

7.3 Toxicity of Anthocyanins

No adverse health effects have been associated with anthocyanins consumption [20, 21]. Even if consumption in large quantities in countries of Europe and USA, no toxicity has been observed, probably due to their low overall absorption. However the over-increasing usage as supplements may prove problematic. In toxicity studies in animal models, using bilberry extracts (36 % anthocyanindins), LD_{50} values were over 2000 mg/kg without any toxic indications [22]. Similarly in dogs, no mortality or toxic effects were observed at 3000 mg/kg. No adverse or toxic effects were observed in rats (@ 125–500 mg/kg) and dogs (@80–320 mg/kg) daily for 6 months. The extract did not exhibit any teratogenic or mutagenic effects thus confirming clinical safety studies [22]. People consuming 160 mg/kg twice daily for 1–2 month tolerated the extract very-well. Only some (4 %) complained side effects, mostly gastrointestinal (GIT) and nervous system and cutaneous annexes and skin [23].

Anthocyanins have been consumed since ancient times but no toxic effects regarding health were reported [24]. That is the reason that Europe, Japan, USA and many other countries allowed the use of anthocyanins as food color in foods and beverages [25]. The Joint FAO/WHO Expert Committee on Food Additives (JECFA) determined on early toxicological studies that anthocyanin-containing extracts had a very low toxicity [26]. The no-observed-effect-level (NOEL) for young rats was determined to be approximately 225 mg kg^{-1} body weight in a two-generation reproduction study. Based on the above result, the estimated acceptable daily intake (ADI) for human was estimated to be 2.5 mg kg^{-1} body weight per day in 1982, using the equation of ADI = NOEL/100 [27]. The studies presented in Tables 7.1 and 7.2 have confirmed the safety of anthocyanins.

Table 7.2 Various mutagenicity studies are described by JECFA [28]

Anthocyanins	Model tested	Genotoxicity	References
Cyanidin and delphinidin	Ames assay system for five different strains of *Salmonella typhimurium*	Not mutagenic	[29]
Cyanidin	Ames assay using *Salmonella typhimurium*strain TA98	Not mutagenic	[30]
Anthocyanins (compounds not specified)	Ames test using *Salmonella typhimurium*TA1538	Not mutagenic	[31]
Anthocyanins (compounds not specified)	*Escherichia coli* WP2 for induction of DNA damage	Not mutagenic	[31]

7.4 Reproductive and Developmental Toxicity

Grape-skin extract containing 3 % anthocyanins were evaluated for 2-generation reproduction study in Sprague–Dawley rats confirmed its safety in reproduction and growth; there were no corresponding compound-related histopathological effects [32]. Anthocyanins (an extract from currants, blueberries and elderberries) were reported not to be a developmental toxicant in rats, mice or rabbits when given over three successive generations [28]. Table 7.3 shows effects of processing conditions on anthocyanin contents (mg/g DW) of fruits.

7.5 Anthocyanins Biological and Pharmacological Activities

Anthocyanins are not only beneficial to its producers, i.e. plants but also to humans and other animals as nutraceutical and coloraceutical. In present era numerous scientists have focused anthocyanins, most of them have found anthocyanins as useful and beneficial reporting wide range of biological activities [11]. These activities include anti-inflammatory [34] antimicrobial [35] anti-carcinogenic [11, 36] vision-improvement [37], stimulation of apoptosis [36]; neuroprotective effects [38, 39] effects on blood vessels [40, 41] platelets [42] and decrease the risk of coronary heart disease [43].

Table 7.3 Effects of processing conditions on anthocyanin contents (mg/g DW) of fruits [33]

	Fresh		Heating (98 °C, 10 min)		Freezing (20 °C)		Freeze-drying	
Fruits grower	Cyd-3-glu	Cyd-3-rut	Cyd-3-glu	Cyd-3-rut	Cyd-3-glu	Cyd-3-rut	Cyd-3-glu	Cyd-3-rut
Cherries								
Cromwell	207	276	419	475	570	737	158	211
Roxburgh	546	698	470	615	548	706	522	628
Dunedin	46	49	126	172	62	89	22	28
Nectarines								
Cromwell	24	25	27	18	33	29	58	7
Roxburgh	16	16	12	12	19	18	71	9
Dunedin	10	15	5	4	9	7	69	7
Peaches								
Cromwell	7	7	12	10	17	18	10	12
Roxburgh	16	17	32	27	33	29	6	8
Dunedin	14	15	10	9	14	8	11	15
Plums								
Cromwell	51	52	38	40	58	63	14	18
Roxburgh	12	13	37	46	46	59	7	8
Dunedin	18	23	48	60	66	85	19	24

Cyd-3-glu cyaniding-3-glucoside, *Cyd-3-rut* cyaniding-3-rutinoside

The purposed mechanism of actions include antioxidant effects, modulation of hormonal and enzyme system, stimulation of the immune system and their interrelated mechanisms. As mentioned in Chap. 5, anthocyanins show less bioavailability, due to their instability under physiological and pH conditions at absorption site in human body, it was difficult to attest their tagged health-promoting effects in animal models in vivo. Although mechanism of action and various health-promoting effects need further investigation, anthocyanins certainly have a key role in a healthy and well-being oriented diet.

7.6 Antioxidant Activity

Anthocyanins exhibit excellent in vitro and in vivo antioxidant activity protecting human body against oxidative stress by scavenging free radicals and suppressing lipid peroxidation. Despite various limitations, much progress has been done during last decade especially in assessing the dynamics of the interplay of ROS and transcription. However the full extent to which anthocyanin is functionally significant in maintaining the cellular redox homeostasis still remains to be understood. The core strategy for improving immune suppression and slowing-down the progression of associated diseases is decreasing the extra production of ROS.

The structure activity relationships of anthocyanins may be considered through various angles like the position and numbers of hydroxyl and methyl groups in anthocyanins nucleus. For example the increase antioxidant effect and higher pro-apoptotic activity in human leukemia cells is considered to be associated with greater number of hydroxyl group. Similarly the antiproliferative activity is thought to be more potent with the hydroxyl or methyl groups substitution or placement on ring B. The phenolic structure of anthocynains imparts antioxidant property to them. Their antioxidant activity depends on many factors mainly; the number of OH group; the catechol moiety in B-ring and oxonium ion in C-ring, the methylation and hydroxylation pattern as well as glycosylation and acylation [44]. Glycosylation decreases antioxidant activity as it decreases the capacity of anthocyanin to delocalize electrons [45]. The contribution of B-ring substituition in antioxidant activity is $OH > OCH3 \gg H$ [46]. The positively charged O-atom in anthocyanin makes it a potential H-donating antioxidant [11].

Anthocyanins can express their antioxidant effects directly or indirectly. Directly they exhibit antioxidant activity by direct free radical scavenging potential due to electron/hydrogen donating capacity of anthocyanin structure [47, 48] which can bind reactive free radical species. Indirectly, anthocyanins can (i) restore or increase the activities of glutathione peroxidase or superoxide dismutase thus increasing glutathione content [49], (ii) activate genes that code for these enzymes [50] and (iii) decrease the generation of oxidative abducts in DNA decreasing the generation of ROS by inhibiting xanthine oxidase and NADH oxidase or by modification of arachidonic metabolism and mitochondrial respiration [1, 51].

The anthocyanidins and anthocyanins have shown a higher antioxidant activity than vitamins C, butylated hydroxyl anisole (BHA), butylated hydroxyl toulene (BHT), α-tocopherol and Vitamin-E [48, 52] by capturing free radicals through the donation of hydrogen atoms [53, 54].

Anthocyanidins have greater antioxidant capacity than anthocyanins i.e. the radical scavenging activity decreases with glycosylation [48]. However, Kähkönen and Heinonen [55] concluded that in vitro effect of glycosylation on antioxidant activity is dependent on the environment e.g. anthocyanidin and oxidation models in which oxidations is occurring [56].

The antioxidant ability of anthocyanins depends on the chemical structure of a anthocyanins. It is not necessary that all of them possess similar activities for scavenging different ROS and reactive nitrogen species (RNS) [56, 57] for example the reactivity against superoxide anion follows the order delphinidine > cyaniding > pelargonidin while pelargonidin is the most effective against the hydroxyl radical [58, 59].

Number and position of sugar residues in the anthocyanidin also affects the antioxidant activity of anthocyanins [55, 59]. The smaller the number of sugar units at C3, higher is the antioxidant potential however the effect is method dependent [55, 60]. Pyranoanthocyanins of malvidin, petunidin, pelargonidin and cyanidin exhibited higher superoxide anion radicals (O^{2-}) quenching activity but did not quench hydroxyl radicals [61], however integration of pyruvic acid into malvidin-3-monoglucoside and delphinidin-3-monoglucoside decreased significantly the antioxidant activity in aqueous-phase systems as suggested by some reports [57, 60].

Greater the number of free OH around the pyrone ring, greater is antioxidant activity. Anthocyanins with 3′, 4′-dihydroxy groups are good chelating agent for metal ions [62]. The chelating capacity also increases at pH 2–4, anthocyanins exist as flavylium cations and due to charge distribution, these are vulnerable to nucleophilic attack on 2 and 4 positions [63]. Not only the hydroxyl group but the position and degree of methoxyl groups on pyrone ring influenced antioxidant potential of anthocyanins [60]. For example cyanidin-3-rutinoside and delphinidin-3-glucoside have greater antioxidant capacity than malvidin-3-glucoside and petunidin-3-glucoside [55, 57].

The antioxidant potential of berries like red raspberries, black raspberries, strawberries and black berries, is directly proportional to the anthocyanins content and has strong scavenging capacity against chemically generated reactive oxygen species [64–66]. Various drinks like whiskeys, sake, Cavas, Chilean Cabernet Sauvignon red wine, port wine or by-products such as grape pomace and fruit and vegetable juices have been reported for antioxidant activity due to presence of anthocyanins [67–71] and a reason behind the French paradox. Red wine fractions having highest anthocyanins substantially decreased reactive oxygen species (ROS) in human red blood cells treated with H_2O_2 as in vitro oxidative model [72]. Recent days the customers expect natural pigment in wine, food and pharmaceutical industry with antioxidant potentials as well [73].

Wang and Mazza reported that cyaniding, pelargonidin, peonidin, malvidin, delphinidin, malvidin-3,5-diglucoside and malviding-3-glucoside possess strong inhibitory effects >50 % comparable quercetin on NO production in LPS/IFN-γ-activated

RAW 264.7 macrophage without showing any cytotoxicity at the range of 16–500 μM [34]. Anthocyanin-rich berry extracts inhibited NO-production and inhibitory effects are directly related to the content of total anthocyanins. The antioxidant capacity of serum increases with the consumption of red wine, strawberries and blueberries the report was confirmed by measuring antioxidant capacity as ORAC, TEAC and TRAP [74, 75].

Anthocyanins are powerful antioxidants in vitro. Using the oxygen radical absorbance capacity (ORAC) assay, antioxidant potential of 14 anthocyanins including delphinidin, cyanidin, pelargonidin, malvidin, peonidin, and their glycosylated derivatives was measured in aqueous phase at neutral pH [5]. Cyanidin-3-glucoside was found to the highest antioxidant (ORAC value, 3.5 times as potent as Trolox) and pg exhibited the lowest ORAC value.

In linoleic acid autoxidation, liposome rabbit erythrocyte membrane, and rat liver microsomal systems, cyanidin-3-glucoside and its aglycone cyanidin were shown to have similar antioxidant potency as vitamin E (α-tocopherol) [76].

Cyanidin-3-glucoside diminished powerfully the changes of biomarkers in hepatic injury in vivo (rat model) hepatic ischemia-reperfusion as an oxidative stress models thus revealing protective effect of anthocyanins on oxidative stress–induced damage [77]. Ramirez-Tortosa et al. found that anthocyanins enhanced plasma antioxidant potential and reduced the level of 8-oxo-deoxyguanosine and hydroperoxides in study of 12 week interval on rats, feeding them vitamin E–deficient diets for followed by supplementation with purified anthocyanin-rich extracts thus indicating significant reductions of the vitamin E deficiency–induced lipid peroxidation and DNA damage [78]. The plasma antioxidant status of Dahl–SS rats, measured by a total antioxidant status (TAS) assay, was restored by a diet supplemented with 1 % freeze-dried whole tart cherry powder for 90 days [79]. Various antioxidant assays used for assessment of antioxidant potential of anthocyanins are shown in Table 7.4.

Tyrosine nitration activity of anthocyanins declined as: cyanidin-3-rutinoside > malvidin-3-monoglucoside ≈ delphinidin-3-monoglucoside > petunidin-3-monoglucoside [60]. Antioxidant activity is dependent on pH and anthocyanins isomeric forms [77]. The peroxynitrite ($ONOO^-$) quenching activity of anthocyanins at pH 7.4 declined as: cyanidin-3-rutinoside > malvidin-3-monoglucoside ≈ delphinidin-3-mono-glucoside > petunidin-3-monoglucoside [60].

7.7 Protection Against Cardiovascular Diseases

The development of cardiovascular disease is mainly due to hypertension, platelet aggregation, dysfunctioning of vascular endothelium and high plasma LDL cholesterol. Anthocyanin exerted cardio-protection under ischemia-perfusion condition besides lowering the oxidative stress in vascular endothelium [97]. Youdim et al. (2000) have proved that anthocyanins are incorporated into the cytosol and membrane of vascular endothelial cells thus protecting against oxidative stress and protecting endothelial function thereby halting the vascular diseases [98]. In a study reporting direct effects of bilberry anthocyanins on whole rat hearts under

Table 7.4 Various antioxidant assays reported for anthocyanins

Name of antioxidant assay	Reference
Oxygen radical absorbance capacity (ORAC)	[3, 5, 65, 74, 75, 80, 81]
Hydrogen transfer-based assay	[3, 5, 65, 80, 82]
Ferric reducing antioxidant potential (FRAP)	[83–88]
Trolox equivalent antioxidant capacity (TEAC)	[74, 75, 81, 83–88]
2,2-diphenyl-1-picrylhydrazyl (DPPH) free radical scavenging activity	[83–88]
Electron transfer-based assays	[83–88]
Superoxide scavenging assay	[89]
Peroxynitrite (ONOO–) scavenging activity	[90]
Inhibition of human low-density lipoprotein	[64]
Inhibition of lipid peroxidation	[90]
Induction of antioxidant enzymes assay e.g. gluthatione-*S*-transferase (GST), gluthationereductase (GR), gluthationeperoxidise (GPx) and superoxide dismutase	[91, 92]
Heavy metal binding assay e.g. iron, zinc and copper	[93]
NO assay	[94]
TRAP	[74, 75, 81]
Human red blood cells treated with H_2O_2 as in vitro oxidative model	[95]
Hepatic ischemia-reperfusion as an oxidative stress model	[77]
Vitamin E deficiency–induced lipid peroxidation	[78]
Serum antioxidant capacity (SAOC)	[96]
Total antioxidant status (TAS) assay	[79]

ischemia-reperfusion situation, the results indicated that perfusion with small amounts of bilberry anthocyanins (0.01–1 mg/L) substantially reduced ischemia-perfusion injury by increasing post-ischemic coronary flow, decreasing the rate of lactate dehydrogenase as well as chances and length of reperfusion arrhythmias [99]. Anaerobic metabolism can decrease intracellular pH under reperfusion conditions thus affecting radical-quenching activities of anthocyanins [100]. All these results suggest that anthocyanin despite their low bioavailability act as cardio-protectants. In in vivo hamster animal model of ischemia-reperfusion, bilberry anthocyanin decreased micro-vascualr injuries by preserving endothelium and enhanced capillary perfusion [101]. In 8 week dietary ingestion of plant-derived anthocyanin in rats, made myocardium less vulnerable to ischemia-reperfusion impairments in vivo as well as ex vivo [49]. Figure 7.2 indicates the relationship between consumption of anthocyanins and their effects on cardiovascular health.

As mentioned above, over production of free radicals and reactive oxygen species (ROS) are the key elements that led to activation of pro-inflammatory mediators that ultimately led to cardiovascular disorders like ischemic heart problem, cardiomyopathy and coronary heart problems. Anthocyanins are most consumed subgroup of flavonoids that control/reduce this heart associated chronic abnormalities via antioxidant, anti-inflammatory and free radical scavenging mechanisms.

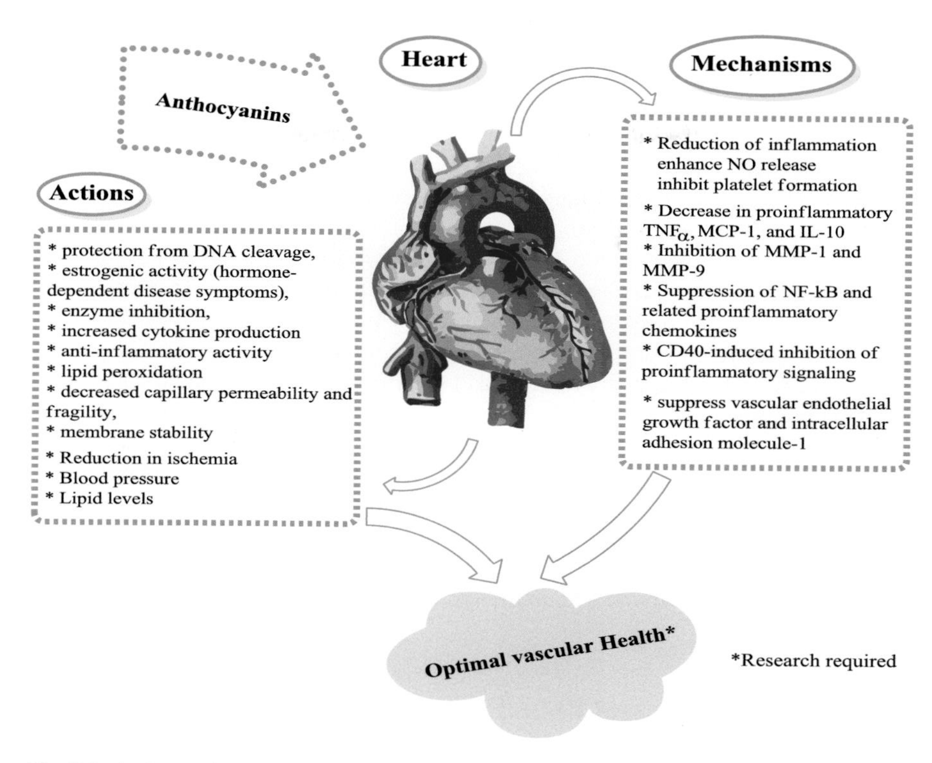

Fig. 7.2 Anthocyanins and cardiovascular health (actions and mechanisms) [102]

Despite the high fat intake there is low incidence of cardiovascular disease amongst the French people (French paradox) which was supposed to be due to the daily consumption of red wine, rich in polyphenols like anthocyanins [43, 103]. Fresh, freeze-dried and purified extracts and juice of chokeberries, cranberries, blueberries, and strawberries anthocyanins have verified significant step up in low density lipoprotein (LDL) oxidation, lipid peroxidation, dyslipidemia, glucose metabolism and total plasma antioxidant status in human blood [104, 105].

Mink et al. (2007) reported in a specific epidemiological study that strawberry and blueberry intake caused reduction of mortality due cardiovascular diseases in postmenopausal woman [106]. The effect was considered to be due to presence of anthocyanins, a question was raised that other flavonoids may also contribute but later in another epidemiological study, Cassidy et al. reported that high consumption of anthocyanins can decrease MI risk in predominantly young women in a comparison of those with lower intake. They confirmed that other flavonoids are not associated with such effects in their study [107] however further details trials are required to note the impact of anthocyanins enriched food.

According to several epidemiological studies, coronary heart disease mortality may be reduced by reasonable intake of red wine [108]. The reduction in mortality due to coronary heart disease may be due to reduction in platelet coagulation [43], increase in circulatory high-density lipoprotein (HDL) [108], free radical scavenging, modulation of eicosanoid metabolism and inhibition of lipoprotein oxidation [109]. The last three mechanisms are also considered to be responsible for reduction of atherosclerosis [75]. Anthocyanins are responsible for reduction in coronary heart diseases and other health beneficial effects so complete metabolomic profile is required in properly designed long term intervention studies. This area has much space to be investigated [75].

Daily intake of concentrated red grape juice (@125 mL) increase serum antioxidant potential thus reducing susceptibility of LDL to oxidation. Thus red grape juice was proposed to possess beneficial effects equal to red wine [110]. Similarly other anthocyanins rich foods e.g. black currant consumption cause a rapid growth in plasma antioxidant capacity until 2 h [111]. Abuja et al. investigated that spray-dried elderberry juice comprising greater anthocyanins content has shielding effect on human LDL in vitro [112]. Tsuda et al. (1996) demonstrate that purified anthocyanins like pelagonidin-3-glucoside, cyanidin-3-glucoside, delphenidin-3-glucoside and their aglycones in a UV-induced lipid peroxidation model, showed strong inhibition of lipid peroxidation [58]. Even anthocyanin colonic metabolite, i.e. protocatechuic acid improved atherosclerosis progression by acting as anti-inflammatory agent and also exhibited antiplatelet activity [113, 114].

7.8 Anthocyanins and Bone Health

Bone health is one of the prime concerns in human health specially aged people; one of the most common occurring bone diseases is osteoporosis. Osteoporosis is characterized by decrease in bone mineral content including decrease in calcium content; age related factors like hormonal imbalance, chronic inflammation and increase in oxidative stress are the main causes [115]. All these factors are linked to reactive oxygen species and to prevent bone loss, direct or indirect control of reactive oxygen species is required. Anthocyanins containing foods are thought to have important role in the prevention of osteoporosis. Different studies reported the confirmed the role of anthocyanins in bone diseases prevention (Fig. 7.3).

Kaume et al. (2015) carried out a study in 9 months old female Sprague–Dawley rats, it was confirmed that feeding blackberries to rats prevent bone desorption especially in ovariectomized rat models, the author observed the effect in dose dependent manner, cyanidin 3-*O*-*β*-D-glucoside may be responsible for the bone loss preventive effect [116]. Women who consumed fruit in greater quantity, in childhood have high bone mineral density compared to normal, similar is the case with all women and men who are consumer of fruit, vegetables and grains [115, 117]. Welch et al. (2013) carried out an observational study in women that were habitual to flavonoids and anthocyanins intake, they observed increased bone mineral density [118]. Similarly the decrease risk of fracture was observed by Langestmo et al. (2011) in individuals that were frequents consumers of fruit, vegetables and grains [119].

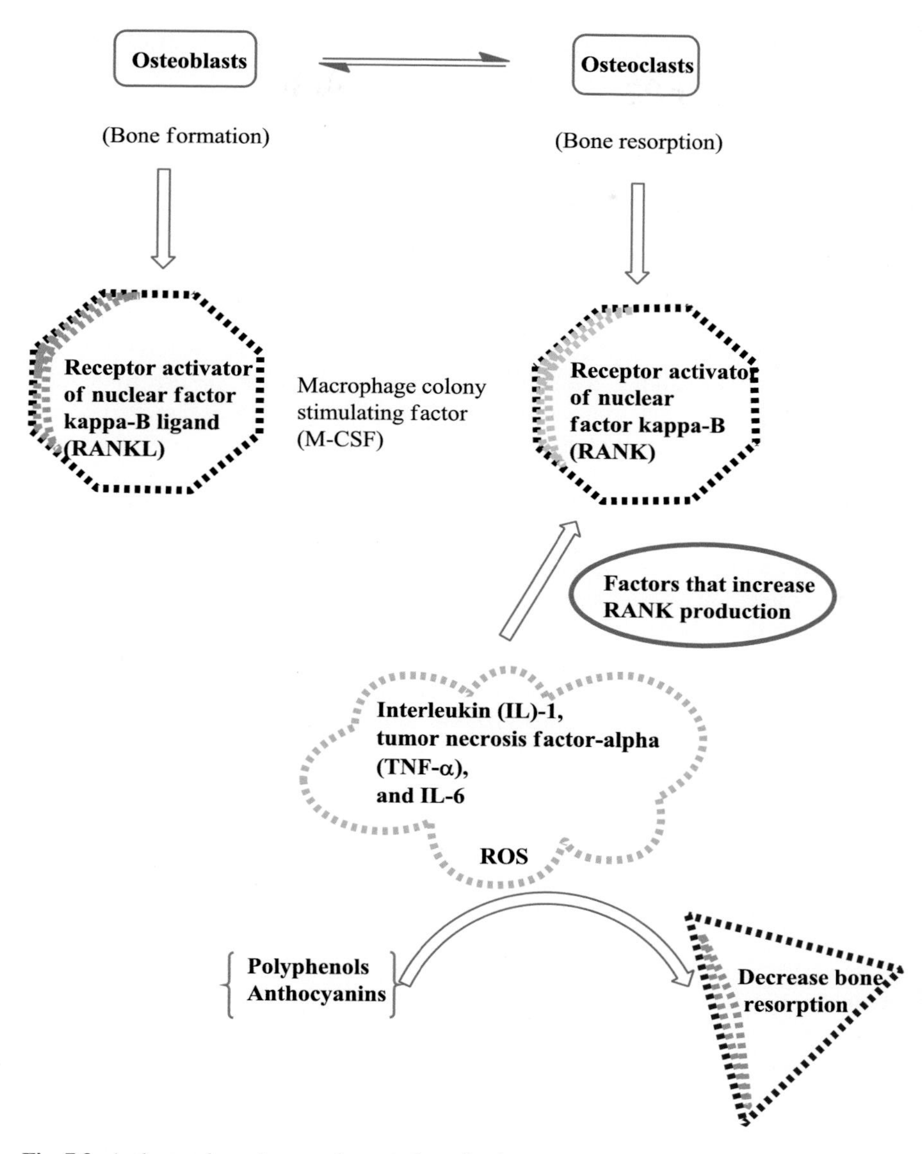

Fig. 7.3 Anthocyanins osteoporosis control mechanism

Tanabe et al. (2011) applied cranberry extract on human bone marrow cells (pre-osteoclastic) at a dose of 10, 25, 50, 100 μg/mL for 4 days duration, the rate of bone degradation by inhibition of RANKL dependent osteoclasts decreased [120]. 500 ng/mL blueberry extract for 72 h duration increased proliferation of human bone marrow cells (CD34+ or CD133) while decreased tartrate resistant acid phosphatase (TRAP) staining and RANKL dependent osteoclast numbers [121]. Six months old SHAM and OVX (ovariectomized) female rats, were used in this study, they were treated in three groups as (i) SHAM Control (ii) OVX-Control and (iii) OVX+5 % blueberry for 100 day duration, it was found that OVX+5 % blueberry group increased whole body bone mineral density (BMD) and serum alkaline phosphatase (ALP) [122].

Chen et al. (2010) and Zhang et al. (2011) studied the affect of blueberry enriched diet in Sprague–Dawley rats and it was found that bone mass and bone mineral density increases, Early blueberry supplementation prevented osteoblast senescence and adult bone loss [123, 124].

Berries are rich source of anthocyanins so the bone loss preventive effect are considered to associated with the antioxidant, and anti-inflammatory effect of anthocyanins however as berries also contain vitamin C so it's could not be excluded.

Dou et al. (2014) studied the effect of cyanidin on bone for the first time, they used RAW264.7 cells and primary bone monocytes in their study design, they found that cyanidin has dual effect on RANKL induced osteoclastogenesis, dose less than 1 μg has promoting while greater than 10 μg has inhibitory effect. Fusogenic and osteoclast related genes are regulated by cyanidin by the same dual pattern thus osteoclast related bone disorders may be controlled with the use of cyanidin [125]. Related group members of anthocyanins delphinidins have been shown to support bone mineralization by suppressing the differentiation and function of osteoclasts, the cells resorbing bone, this statement is supported by study in ovariectomized mice model through oral administration of delphinidin [126, 127]. These investigations suggest a potential benefit from regular consumption of anthocyanins rich sources for preventing bone demineralization at higher age.

7.9 Conclusions

Despite all above mentioned effects, it is occasionally difficult to discriminate the effects of anthocyanins from other constituents or even isolation of effect of a single anthocyanin molecule from total pool. Further the doses mentioned in literature for animal models are sometimes far from those compatible with realistic values for human body due to bioavailability differences in both. There are enticing glimpses of health-promoting effects however many questions remain unanswered since most of these studies are in vitro. There is need of balanced, data-driven answers for questions raised by nutritionists and health professionals regarding the totality of evidence rather than findings of any single study. There is still unceratininity about some mechanisms. There is plenty of information about role of anthocyanins in prevention of major diseases however many tropical and emerging diseases remain un-attentioned. Considering complex biological functions and their efficacy in prevention of various chronic human diseases, combination therapies with other natural antioxidants can be a useful approach for various diseases.

References

1. Pojer, E., Mattivi, F., Johnson, D., & Stockley, C. S. (2013). The case for anthocyanin consumption to promote human health: A review. *Comprehensive Reviews in Food Science and Food Safety, 12*(5), 483–508.
2. Timberlake, C. (1988). The biological properties of anthocyanin compounds. *NATCOL Quarterly Bulletin, 1*, 4–15.
3. Prior, R. L., Cao, G., Martin, A., Sofic, E., McEwen, J., O'Brien, C., et al. (1998). Antioxidant capacity as influenced by total phenolic and anthocyanin content, maturity, and variety of *Vaccinium* species. *Journal of Agricultural and Food Chemistry, 46*(7), 2686–2693.
4. Degenhardt, A., Knapp, H., & Winterhalter, P. (2000). Separation and purification of anthocyanins by high-speed countercurrent chromatography and screening for antioxidant activity. *Journal of Agricultural and Food Chemistry, 48*(2), 338–343.
5. Wang, H., Cao, G., & Prior, R. L. (1997). Oxygen radical absorbing capacity of anthocyanins. *Journal of Agricultural and Food Chemistry, 45*(2), 304–309.
6. Lietti, A., Cristoni, A., & Picci, M. (1976). Studies on *Vaccinium myrtillus* anthocyanosides. I. Vasoprotective and anti-inflamatory activity. *Arzneimittel-Forschung, 26*(5), 829–832.
7. Morazzoni, P., & Magistretti, M. (1986). Effects of *Vaccinium myrtillus* anthocyanosides on prostacyclin-like activity in rat arterial tissue. *Fitoterapia, 57*, 11–14.
8. Kamei, H., Kojima, T., Hasegawa, M., Koide, T., Umeda, T., Yukawa, T., et al. (1995). Suppression of tumor cell growth by anthocyanins in vitro. *Cancer Investigation, 13*(6), 590–594.
9. Karaivanova, M., Drenska, D., & Ovcharov, R. (1990). A modification of the toxic effects of platinum complexes with antocyans. *Eksperimentalna Meditsina i Morfologiia, 29*(2), 19.
10. Akhmadieva, A. K., Zaichkina, S. I., Ruzieva, R. K., & Ganassi, E. E. (1993). The protective action of a natural preparation of anthocyan (pelargonidin-3, 5-diglucoside). *Radio Biologica, 33*(3), 433.
11. Kong, J.-M., Chia, L. S., Goh, N. K., Chia, T. F., & Brouillard, R. (2003). Analysis and biological activities of anthocyanins. *Phytochemistry, 64*(5), 923–933.
12. Valls, J., Millán, S., Martí, M. P., Borràs, E., & Arola, L. (2009). Advanced separation methods of food anthocyanins, isoflavones and flavanols. *Journal of Chromatography A, 1216*(43), 7143–7172.
13. Cardile, V., Frasca, G., Rizza, L., Rapisarda, P., & Bonina, F. (2010). Anti-inflammatory effects of a red orange extract in human keratinocytes treated with interferon gamma and histamine. *Phytotherapy Research, 24*(3), 414–418.
14. Lorencini, M., Brohem, C. A., Dieamant, G. C., Zanchin, N. I., & Maibach, H. I. (2014). Active ingredients against human epidermal aging. *Ageing Research Reviews, 15*, 100–115.
15. Mantena, S. K., & Katiyar, S. K. (2006). Grape seed proanthocyanidins inhibit UV-radiation-induced oxidative stress and activation of MAPK and NF-κB signaling in human epidermal keratinocytes. *Free Radical Biology and Medicine, 40*(9), 1603–1614.
16. Scalbert, A., & Williamson, G. (2000). Dietary intake and bioavailability of polyphenols. *Journal of Nutrition, 130*(8), 2073S–2085S.
17. Prior, R. L., & Wu, X. (2006). Anthocyanins: Structural characteristics that result in unique metabolic patterns and biological activities. *Free Radical Research, 40*(10), 1014–1028.
18. Wang, S. Y., & Jiao, H. (2000). Scavenging capacity of berry crops on superoxide radicals, hydrogen peroxide, hydroxyl radicals, and singlet oxygen. *Journal of Agricultural and Food Chemistry, 48*(11), 5677–5684.
19. Shih, P.-H., & Yen, G.-C. (2007). Differential expressions of antioxidant status in aging rats: The role of transcriptional factor Nrf2 and MAPK signaling pathway. *Biogerontology, 8*(2), 71–80.
20. Brouillard, R. (1982). *Chemical structure of anthocyanins* (Vol. 1). New York: Academic.
21. Markakis, P. (2012). *Anthocyanins as food colors*. New York: Elsevier.
22. Morazzoni, P., & Bombardelli, E. (1996). *Vaccinium myrtillus* L. *Fitoterapia, 67*(1), 3–29.

23. He, J., & Giusti, M. M. (2010). Anthocyanins: Natural colorants with health-promoting properties. *Annual Review of Food Science and Technology, 1*, 163–187.
24. Brouillard, R., & Markakis, P. (1982). *Anthocyanins as food colors*. New York: Academic.
25. Eder, R., & Nollet, L. (2000). Pigments. In M. L. L. Nollet (Ed.), *Food analysis by HPLC* (pp. 825–880). New York: Marcer Dekker.
26. Organization, W. H. (1982). Anthocyanins. *Toxicological Evaluation of Food Additives: Technical Report Series, 17*, 42–49.
27. Clifford, M. N. (2000). Anthocyanins-nature, occurrence and dietary burden. *Journal of the Science of Food and Agriculture, 80*(7), 1063–1072.
28. EFSA Panel on Food Additives and Nutrient Sources added to Food. (2013). Statement on two reports published after the closing date of the public consultation of the draft Scientific Opinion on the re-evaluation of aspartame (E 951) as a food additive. *European Food Safety Authority Journal, 11*(12), 3504.
29. Brown, J. P., & Dietrich, P. S. (1979). Mutagenicity of plant flavonols in the Salmonella/mammalian microsome test: Activation of flavonol glycosides by mixed glycosidases from rat cecal bacteria and other sources. *Mutation Research, 66*(3), 223–240.
30. Macgregor, J. T., & Jurd, L. (1978). Mutagenicity of plant flavonoids: Structural requirements for mutagenic activity in *Salmonella typhimurium*. *Mutation Research, 54*(3), 297–309.
31. Haveland-Smith, R. (1981). Evaluation of the genotoxicity of some natural food colours using bacterial assays. *Mutation Research Letters, 91*(4), 285–290.
32. Cox, G., Rucci, G., & Babish, J. (1978). *90-day subacute dietary toxicity study of 78-002-2 in Sprague-Dawley rats*. Unpublished report submitted to the Flavor and Extract Manufacturers Association, Washington, DC, USA, by Food and Drug Research Laboratories, Inc. Submitted to WHO by the International Organization of the Flavor Industry, Brussels, Belgium.
33. Leong, S. Y., & Oey, I. (2012). Effects of processing on anthocyanins, carotenoids and vitamin C in summer fruits and vegetables. *Food Chemistry, 133*(4), 1577–1587.
34. Wang, J., & Mazza, G. (2002). Effects of anthocyanins and other phenolic compounds on the production of tumor necrosis factor α in LPS/IFN-γ-activated RAW 264.7 macrophages. *Journal of Agricultural and Food Chemistry, 50*(15), 4183–4189.
35. Pisha, E., & Pezzuto, J. (1994). Fruits and vegetables containing compounds that demonstrate pharmacological activity in humans. *Economic and Medicinal Plant Research, 6*, 189–233.
36. Katsube, N., Iwashita, K., Tsushida, T., Yamaki, K., & Kobori, M. (2003). Induction of apoptosis in cancer cells by bilberry (Vaccinium myrtillus) and the anthocyanins. *Journal of Agricultural and Food Chemistry, 51*(1), 68–75.
37. Matsumoto, H., Nakamura, Y., Iida, H., Ito, K., & Ohguro, H. (2006). Comparative assessment of distribution of blackcurrant anthocyanins in rabbit and rat ocular tissues. *Experimental Eye Research, 83*(2), 348–356.
38. Youdim, K., Shukitt-Hale, B., MacKinnon, S., Kalt, W., & Joseph, J. A. (2000). Polyphenolics enhance red blood cell resistance to oxidative stress: In vitro and in vivo. *Biochimica et Biophysica Acta, 1523*(1), 117–122.
39. Galli, R. L., Shukitt-Hale, B., Youdim, K. A., & Joseph, J. A. (2002). Fruit polyphenolics and brain aging: nutritional interventions targeting age-related neuronal and behavioral deficits. *Annals of the New York Academy of Science, 959*(1), 128–132.
40. Andriambeloson, E., Magnier, C., Haan-Archipoff, G., Lobstein, A., Anton, R., Beretz, A., et al. (1998). Natural dietary polyphenolic compounds cause endothelium-dependent vasorelaxation in rat thoracic aorta. *Journal of Nutrition, 128*(12), 2324–2333.
41. Martin, P. D., Warwick, M. J., Dane, A. L., Brindley, C., & Short, T. (2003). Absolute oral bioavailability of rosuvastatin in healthy white adult male volunteers. *Clinical Therapeutics, 25*(10), 2553–2563.
42. Demrow, H. S., Slane, P. R., & Folts, J. D. (1995). Administration of wine and grape juice inhibits in vivo platelet activity and thrombosis in stenosed canine coronary arteries. *Circulation, 91*(4), 1182–1188.

43. Renaud, S. D., & de Lorgeril, M. (1992). Wine, alcohol, platelets, and the French paradox for coronary heart disease. *Lancet, 339*(8808), 1523–1526.
44. Yang, M., Koo, S. I., Song, W. O., & Chun, O. K. (2011). Food matrix affecting anthocyanin bioavailability: Review. *Current Medicinal Chemistry, 18*(2), 291–300.
45. Wang, L.-S., & Stoner, G. D. (2008). Anthocyanins and their role in cancer prevention. *Cancer Letters, 269*(2), 281–290.
46. Rossetto, M., Vanzani, P., Lunelli, M., Scarpa, M., Mattivi, F., & Rigo, A. (2007). Peroxyl radical trapping activity of anthocyanins and generation of free radical intermediates. *Free Radical Research, 41*(7), 854–859.
47. Borkowski, T., Szymusiak, H., Gliszczyńska-Rwigło, A., Rietjens, I. M., & Tyrakowska, B. (2005). Radical scavenging capacity of wine anthocyanins is strongly pH-dependent. *Journal of Agricultural and Food Chemistry, 53*(14), 5526–5534.
48. Fukumoto, L., & Mazza, G. (2000). Assessing antioxidant and prooxidant activities of phenolic compounds. *Journal of Agricultural and Food Chemistry, 48*(8), 3597–3604.
49. Toufektsian, M.-C., de Lorgeril, M., Nagy, N., Salen, P., Donati, M. B., Giordano, L., et al. (2008). Chronic dietary intake of plant-derived anthocyanins protects the rat heart against ischemia-reperfusion injury. *Journal of Nutrition, 138*(4), 747–752.
50. Shih, P.-H., Yeh, C.-T., & Yen, G.-C. (2005). Effects of anthocyanidin on the inhibition of proliferation and induction of apoptosis in human gastric adenocarcinoma cells. *Food and Chemical Toxicology, 43*(10), 1557–1566.
51. Steffen, Y., Gruber, C., Schewe, T., & Sies, H. (2008). Mono-O-methylated flavanols and other flavonoids as inhibitors of endothelial NADPH oxidase. *Archives of Biochemistry and Biophysics, 469*(2), 209–219.
52. Bagchi, D., Garg, A., Krohn, R. L., Bagchi, M., Bagchi, D. J., Balmoori, J., et al. (1998). Protective effects of grape seed proanthocyanidins and selected antioxidants against TPA-induced hepatic and brain lipid peroxidation and DNA fragmentation, and peritoneal macrophage activation in mice. *General Pharmacology, 30*(5), 771–776.
53. Chen, Z. Y., Chan, P. T., Ho, K. Y., Fung, K. P., & Wang, J. (1996). Antioxidant activity of natural flavonoids is governed by number and location of their aromatic hydroxyl groups. *Chemistry and Physics of Lipids, 79*(2), 157–163.
54. Rice-Evans, C. A., Miller, N. J., & Paganga, G. (1996). Structure-antioxidant activity relationships of flavonoids and phenolic acids. *Free Radical Biology and Medicine, 20*(7), 933–956.
55. Kähkönen, M. P., & Heinonen, M. (2003). Antioxidant activity of anthocyanins and their aglycons. *Journal of Agricultural and Food Chemistry, 51*(3), 628–633.
56. Kay, C. D., Mazza, G., Holub, B. J., & Wang, J. (2004). Anthocyanin metabolites in human urine and serum. *British Journal of Nutrition, 91*(06), 933–942.
57. Miguel, M. (2011). *Anthocyanins: Antioxidant and/or anti-inflammatory activities.*
58. Tsuda, T., Shiga, K., Ohshima, K., Kawakishi, S., & Osawa, T. (1996). Inhibition of lipid peroxidation and the active oxygen radical scavenging effect of anthocyanin pigments isolated from Phaseolus vulgaris L. *Biochemical Pharmacology, 52*(7), 1033–1039.
59. Antal, D.-S., Garban, G., & Garban, Z. (2003). The anthocyans: Biologicallyactive substances of food and pharmaceutic interest. *Annals of the University Dunarea de Jos of Galati, Food Technology, 6*, 106–115.
60. Muselík, J., García-Alonso, M., Martín-López, M. P., Žemlička, M., & Rivas-Gonzalo, J. C. (2007). Measurement of antioxidant activity of wine catechins, procyanidins, anthocyanins and pyranoanthocyanins. *International Journal of Molecular Science, 8*(8), 797–809.
61. Garcia Alonso, M., Rimbach, G., Sasai, M., Nakahara, M., Matsugo, S., Uchida, Y., et al. (2005). Electron spin resonance spectroscopy studies on the free radical scavenging activity of wine anthocyanins and pyranoanthocyanins. *Molecular Nutrition and Food Research, 49*(12), 1112–1119.
62. Sarma, A. D., Sreelakshmi, Y., & Sharma, R. (1997). Antioxidant ability of anthocyanins against ascorbic acid oxidation. *Phytochemistry, 45*(4), 671–674.

63. Bkowska-Barczak, A. (2005). Acylated anthocyanins as stable, natural food colorants – A review. *Polish Journal of Food and Nutrition Sciences, 14*(2), 107–116.
64. Heinonen, I. M., Meyer, A. S., & Frankel, E. N. (1998). Antioxidant activity of berry phenolics on human low-density lipoprotein and liposome oxidation. *Journal of Agricultural and Food Chemistry, 46*(10), 4107–4112.
65. Wang, S. Y., & Lin, H.-S. (2000). Antioxidant activity in fruits and leaves of blackberry, raspberry, and strawberry varies with cultivar and developmental stage. *Journal of Agricultural and Food Chemistry, 48*(2), 140–146.
66. Castañeda-Ovando, A., de Lourdes, P. H. M., Elena, P. H., José, A. R., & Carlos, A. G. V. (2009). Chemical studies of anthocyanins: A review. *Food Chemistry, 113*(4), 859–871.
67. McPhail, D. B., Peter, T. G., Garry, G. D., Gordon, M. S., & Kenneth, R. (1999). Assessment of the antioxidant potential of Scotch whiskeys by electron spin resonance spectroscopy: Relationship to hydroxyl-containing aromatic components. *Journal of Agricultural and Food Chemistry, 47*(5), 1937–1941.
68. Kitagaki, H., & Tsugawa, M. (1999). 1,1-Diphenyl-2-picrylhydrazyl radical (DPPH) scavenging ability of sake during storage. *Journal of Bioscience and Bioengineering, 87*(3), 328–332.
69. Satué-Gracia, M. T., Andrés-Lacueva, C., Lamuela-Raventós, R. M., & Frankel, E. N. (1999). Spanish sparkling wines (cavas) as inhibitors of in vitro human low-density lipoprotein oxidation. *Journal of Agricultural and Food Chemistry, 47*(6), 2198–2202.
70. Aguirre, M. J., Chen, Y. Y., Mauricio, I., Betty, M., Leonora, M., & Simonet, T. (2010). Electrochemical behaviour and antioxidant capacity of anthocyanins from Chilean red wine, grape and raspberry. *Food Chemistry, 121*(1), 44–48.
71. Oliveira, B. F., Nogueira-Machado, J. A., & Chaves, M. M. (2010). The role of oxidative stress in the aging process. *Scientific World Journal, 10*, 1121–1128.
72. Tedesco, I., Luigi Russo, G., Nazzaro, F., Russo, M., & Palumbo, R. (2001). Antioxidant effect of red wine anthocyanins in normal and catalase-inactive human erythrocytes. *Journal of Nutrition and Biochemistry, 12*(9), 505–511.
73. Li, X., Ma, H., Huang, H., Li, D., & Yao, S. (2013). Natural anthocyanins from phytoresources and their chemical researches. *Natural Product Research, 27*(4-5), 456–469.
74. Mazza, G., Kay, C. D., Cottrell, T., & Holub, B. J. (2002). Absorption of anthocyanins from blueberries and serum antioxidant status in human subjects. *Journal of Agricultural and Food Chemistry, 50*(26), 7731–7737.
75. Mazza, G. (2007). Anthocyanins and heart health. *Annali dell'Istituto Superiore di Sanità, 43*(4), 369.
76. Tsuda, T., Mie, W., Katsumi, O., Seiji, N., Sang-Won, C., Shunro, K., et al. (1994). Antioxidative activity of the anthocyanin pigments cyanidin 3-O-β-D-glucoside and cyanidin. *Journal of Agricultural and Food Chemistry, 42*(11), 2407–2410.
77. Tsuda, T., Horio, F., & Osawa, T. (2000). The role of anthocyanins as an antioxidant under oxidative stress in rats. *Biofactors, 13*(14), 133–139.
78. Ramirez-Tortosa, C., Andersen, Ø. M., Cabrita, L., Gardner, P. T., Morrice, P. C., Wood, S. G., et al. (2001). Anthocyanin-rich extract decreases indices of lipid peroxidation and DNA damage in vitamin E-depleted rats. *Free Radical Biology and Medicine, 31*(9), 1033–1037.
79. Seymour, E. M., Singer, A. A., Kirakosyan, A., Urcuyo-Llanes, D. E., Kaufman, P. B., & Bolling, S. F. (2008). Altered hyperlipidemia, hepatic steatosis, and hepatic peroxisome proliferator-activated receptors in rats with intake of tart cherry. *Journal of Medicinal Food, 11*(2), 252–259.
80. Steed, L., & Truong, V. D. (2008). Anthocyanin content, antioxidant activity, and selected physical properties of flowable purple-fleshed sweet potato purees. *Journal of Food Science, 73*(5), S215–S221.
81. Cao, G., Russell, R. M., Lischner, N., & Prior, R. L. (1998). Serum antioxidant capacity is increased by consumption of strawberries, spinach, red wine or vitamin C in elderly women. *Journal of Nutrition, 128*(12), 2383–2390.

82. Zheng, W., & Wang, S. Y. (2003). Oxygen radical absorbing capacity of phenolics in blueberries, cranberries, chokeberries, and lingonberries. *Journal of Agricultural and Food Chemistry, 51*(2), 502–509.
83. Moyer, R. A., Hummer, K. E., Finn, C. E., Frei, B., & Wrolstad, R. E. (2002). Anthocyanins, phenolics, and antioxidant capacity in diverse small fruits: Vaccinium, Rubus, and Ribes. *Journal of Agricultural and Food Chemistry, 50*(3), 519–525.
84. Lee, J., Koo, N., & Min, D. (2004). Reactive oxygen species, aging, and antioxidative nutraceuticals. *Comprehensive Reviews in Food Science and Food Safety, 3*(1), 21–33.
85. Einbond, L. S., Einbond, K. A., Reynertson, X. D. L., Margaret, J. B., & Edward, J. K. (2004). Anthocyanin antioxidants from edible fruits. *Food Chemistry, 84*(1), 23–28.
86. Nakajima, J.-I., Tanaka, I., Seo, S., Yamazaki, M., & Saito, K. (2004). LC/PDA/ESI-MS profiling and radical scavenging activity of anthocyanins in various berries. *BioMed Research International, 2004*(5), 241–247.
87. Solomon, A., Golubowicz, S., Yablowicz, Z., Grossman, S., Bergman, M., Gottlieb, H. E., et al. (2006). Antioxidant activities and anthocyanin content of fresh fruits of common fig (Ficus carica L.). *Journal of Agricultural and Food Chemistry, 54*(20), 7717–7723.
88. Koca, I., & Karadeniz, B. (2009). Antioxidant properties of blackberry and blueberry fruits grown in the Black Sea Region of Turkey. *Scientia Horticulturae, 121*(4), 447–450.
89. Costantino, L., Albasini, A., Rastelli, G., & Benvenuti, S. (1992). Activity of polyphenolic crude extracts as scavengers of superoxide radicals and inhibitors of xanthine oxidase. *Planta Medica, 58*(4), 342–344.
90. Muselík, J., García-Alonso, M., Martín-López, M. P., Žemlička, M., & Rivas-Gonzalo, J. C. (2007). Measurement of antioxidant activity of wine catechins, procyanidins, anthocyanins and pyranoanthocyanins. *International Journal of Molecular Sciences, 8*(8), 797–809.
91. Fiander, H., & Schneider, H. (2000). Dietary ortho phenols that induce glutathione S-transferase and increase the resistance of cells to hydrogen peroxide are potential cancer chemopreventives that act by two mechanisms: The alleviation of oxidative stress and the detoxification of mutagenic xenobiotics. *Cancer Letters, 156*(2), 117–124.
92. Turner, M. K. (2009). *Anthocyanins increase antioxidant enzyme activity in HT-29 adenocarcinoma cells*. MS Thesis, University of Georgia.
93. Havsteen, B. (1983). Flavonoids, a class of natural products of high pharmacological potency. *Biochemical Pharmacology, 32*(7), 1141–1148.
94. Wang, J., & Mazza, G. (2002). Inhibitory effects of anthocyanins and other phenolic compounds on nitric oxide production in LPS/IFN-γ-activated RAW 264.7 macrophages. *Journal of Agricultural and Food Chemistry, 50*(4), 850–857.
95. Tedesco, I., Luigi Russo, G., Nazzaro, F., Russo, M., & Palumbo, R. (2001). Antioxidant effect of red wine anthocyanins in normal and catalase-inactive human erythrocytes. *Journal of Nutritional Biochemistry, 12*(9), 505–511.
96. Whitehead, T. P., Robinson, D., Allaway, S., Syms, J., & Hale, A. (1995). Effect of red wine ingestion on the antioxidant capacity of serum. *Clinical Chemistry, 41*(1), 32–35.
97. Ziberna, L., Tramer, F., Moze, S., Vrhovsek, U., Mattivi, F., & Passamonti, S. (2012). Transport and bioactivity of cyanidin 3-glucoside into the vascular endothelium. *Free Radical Biology and Medicine, 52*(9), 1750–1759.
98. Youdim, K. A., Martin, A., & Joseph, J. A. (2000). Incorporation of the elderberry anthocyanins by endothelial cells increases protection against oxidative stress. *Free Radical Biology and Medicine, 29*(1), 51–60.
99. Halliwell, B. (2008). Are polyphenols antioxidants or pro-oxidants? What do we learn from cell culture and in vivo studies? *Archives of Biochemistry and Biophysics, 476*(2), 107–112.
100. Cvorovic, J., Tramer, F., Granzotto, M., Candussio, L., Decorti, G., & Passamonti, S. (2010). Oxidative stress-based cytotoxicity of delphinidin and cyanidin in colon cancer cells. *Archives of Biochemistry and Biophysics, 501*(1), 151–157.
101. Bertuglia, S., Malandrino, S., & Colantuoni, A. (1995). Effect of *Vaccinium myrtillus* anthocyanosides on ischaemia reperfusion injury in hamster cheek pouch microcirculation. *Pharmacological Research, 31*(3), 183–187.

102. Wallace, T. C. (2011). Anthocyanins in cardiovascular disease. *Advances in Nutrition, 2*(1), 1–7.
103. Kanner, J., Edwin, F., Rina, G., Bruce, G., & John, E. K. (1994). Natural antioxidants in grapes and wines. *Journal of Agricultural and Food Chemistry, 42*(1), 64–69.
104. Basu, A., Rhone, M., & Lyons, T. J. (2010). Berries: Emerging impact on cardiovascular health. *Nutrition Reviews, 68*(3), 168–177.
105. Rodriguez-Mateos, A., Ishisaka, A., Mawatari, K., Vidal-Diez, A., Spencer, J. P., & Terao, J. (2013). Blueberry intervention improves vascular reactivity and lowers blood pressure in high-fat-, high-cholesterol-fed rats. *British Journal of Nutrition, 109*(10), 1746–1754.
106. Mink, P. J., Scrafford, C. G., Barraj, L. M., Harnack, L., Hong, C. P., Nettleton, J. A., et al. (2007). Flavonoid intake and cardiovascular disease mortality: A prospective study in postmenopausal women. *American Journal of Clinical Nutrition, 85*(3), 895–909.
107. Cassidy, A., Mukamal, K. J., Liu, L., Franz, M., Eliassen, A. H., & Rimm, E. B. (2013). High anthocyanin intake is associated with a reduced risk of myocardial infarction in young and middle-aged women. *Circulation, 127*(2), 188–196.
108. Klatsky, A. L., Tekawa, I., Armstrong, M. A., & Sidney, S. (1994). The risk of hospitalization for ischemic heart disease among Asian Americans in northern California. *American Journal of Public Health, 84*(10), 1672–1675.
109. Esterbauer, H., Gebicki, J., Puhl, H., & Jürgens, G. (1992). The role of lipid peroxidation and antioxidants in oxidative modification of LDL. *Free Radical Biology and Medicine, 13*(4), 341–390.
110. Day, A. P., Kemp, H. J., Bolton, C., Hartog, M., & Stansbie, D. (1997). Effect of concentrated red grape juice consumption on serum antioxidant capacity and low-density lipoprotein oxidation. *Annals of Nutrition and Metabolism, 41*(6), 353–357.
111. Matsumoto, H., Nakamura, Y., Hirayama, M., Yoshiki, Y., & Okubo, K. (2002). Antioxidant activity of black currant anthocyanin aglycons and their glycosides measured by chemiluminescence in a neutral pH region and in human plasma. *Journal of Agricultural and Food Chemistry, 50*(18), 5034–5037.
112. Abuja, P. M., Murkovic, M., & Pfannhauser, W. (1998). Antioxidant and prooxidant activities of elderberry (Sambucus nigra) extract in low-density lipoprotein oxidation. *Journal of Agricultural and Food Chemistry, 46*(10), 4091–4096.
113. Rechner, A. R., & Kroner, C. (2005). Anthocyanins and colonic metabolites of dietary polyphenols inhibit platelet function. *Thrombosis Research, 116*(4), 327–334.
114. Wang, D., Zou, T., Yang, Y., Yan, X., & Ling, W. (2011). Cyanidin-3-O-β-glucoside with the aid of its metabolite protocatechuic acid, reduces monocyte infiltration in apolipoprotein E-deficient mice. *Biochemical Pharmacology, 82*(7), 713–719.
115. Hubert, P. A., Lee, S. G., Lee, S. K., & Chun, O. K. (2014). Dietary polyphenols, berries, and age-related bone loss: A review based on human, animal, and cell studies. *Antioxidants, 3*(1), 144–158.
116. Kaume, L., Gilbert, W., Smith, B. J., & Devareddy, L. (2015). Cyanidin 3-O-β-d-glucoside improves bone indices. *Journal of Medicinal Food, 18*(6), 690–697.
117. New, S. A., Robins, S. P., Campbell, M. K., Martin, J. C., Garton, M. J., Bolton-Smith, C., et al. (2000). Dietary influences on bone mass and bone metabolism: Further evidence of a positive link between fruit and vegetable consumption and bone health? *American Journal of Clinical Nutrition, 71*(1), 142–151.
118. Welch, A., MacGregor, A., Jennings, A., Fairweather-Tait, S., Spector, T., & Cassidy, A. (2012). Habitual flavonoid intakes are positively associated with bone mineral density in women. *Journal of Bone and Mineral Research, 27*(9), 1872–1878.
119. Langsetmo, L., Hanley, D. A., Prior, J. C., Barr, S. I., Anastassiades, T., Towheed, T., et al. (2011). Dietary patterns and incident low-trauma fractures in postmenopausal women and men aged $\geq$ 50 y: A population-based cohort study. *American Journal of Clinical Nutrition, 93*(1), 192–199.
120. Tanabe, S., Santos, J., La, V. D., Howell, A. B., & Grenier, D. (2011). A-type cranberry proanthocyanidins inhibit the RANKL-dependent differentiation and function of human osteoclasts. *Molecules, 16*(3), 2365–2374.

121. Bickford, P. C., Tan, J., Shytle, R. D., Sanberg, C. D., El-Badri, N., & Sanberg, P. R. (2006). Nutraceuticals synergistically promote proliferation of human stem cells. *Stem Cells and Development, 15*(1), 118–123.
122. Devareddy, L., Hooshmand, S., Collins, J. K., Lucas, E. A., Chai, S. C., & Arjmandi, B. H. (2008). Blueberry prevents bone loss in ovariectomized rat model of postmenopausal osteoporosis. *Journal of Nutritional Biochemistry, 19*(10), 694–699.
123. Chen, J. R., Lazarenko, O. P., Wu, X., Kang, J., Blackburn, M. L., Shankar, K., et al. (2010). Dietary-induced serum phenolic acids promote bone growth via p38 MAPK/β-catenin canonical Wnt signaling. *Journal of Bone and Mineral Research, 25*(11), 2399–2411.
124. Zhang, J., Lazarenko, O. P., Blackburn, M. L., Shankar, K., Badger, T. M., & Ronis, M. J. (2011). Feeding blueberry diets in early life prevent senescence of osteoblasts and bone loss in ovariectomized adult female rats. *PLoS One, 6*(9), e24486.
125. Dou, C., Li, J., Kang, F., Cao, Z., Yang, X., Jiang, H., et al. (2014). Dual effect of cyanidin on RANKL-induced differentiation and fusion of osteoclasts. *Journal of Cellular Physiology*. doi:10.1002/jcp.24916.
126. Moriwaki, S., Suzuki, K., Muramatsu, M., Nomura, A., Inoue, F., Into, T., et al. (2014). Delphinidin, one of the major anthocyanidins, prevents bone loss through the inhibition of excessive osteoclastogenesis in osteoporosis model mice. *PLoS One, 9*(5), e97177.
127. Watson, R., & Schönlau, F. (2015). Nutraceutical and antioxidant effects of a delphinidin-rich maqui berry extract Delphinol®: A review. *Minerva Cardioangiologica, 63*(2 Suppl 1), 1–12.

Chapter 8
The Role of Anthocyanins in Obesity and Diabetes

8.1 Introduction

Food and phytomedicines are now becoming a trusted remedy and therapy for various types of chronic diseases. The observed interfaces among substances present in a phytochemical mixture strongly support the dogma of 'eating the whole food' rather than purified compounds, extracts, nutraceuticals, or formulation. In case of formulation, phytochemicals can loose interactions during the product development phase leading to decreased potency of extract. As noted in previous chapters, anthocyanins possess a wide spectrum of biological and pharmacological effects, such as antioxidants, anti-cancer, anti-inflammatory, liver support by induction of detoxification enzymes, and induction of apoptosis. As known for other phytochemicals, mixtures of interacting anthocyanins offer boosted synergistic curative and therapeutic potential by multi-pronged pathways of intercession concurrently. Numerous mechanisms of action of anthocyanins at cellular and biochemical levels can be complementary and overlapping, and a blend of these pathways leads to the observed health benefits of ingested anthocyanins. A lot of available studies indicate potentiating interactions between various anthocyanins, and between anthocyanins and other bioactives.

As discussed in Chaps. 1 and 7, many studies suggest anthocyanins as promising molecules for treatment of various diseases such as cancer, diabetes, ulcer, cardiovascular disorders and various metabolic disorders. While observational and lab experimental studies appear to link dietary intake of anthocyanins to improvement in health outcomes, data from randomized controlled trials is limited. In this regard, two factors should be kept in mind, first only small portion of ingested anthocyanins are absorbed to exert any biological effect and second stability of anthocyanins is regulated by various factors. Since most studies describing beneficial effects of anthocyanins are performed in vitro, more studies should be conducted to clarify the mechanism of action as well as dose required to get that effect.

M. Riaz et al., *Anthocyanins and Human Health*, SpringerBriefs in Food, Health, and Nutrition, DOI 10.1007/978-3-319-26456-1_8

8.2 Anthocyanins and Obesity

Obesity is defined as an excessive accumulation of adipose tissue due to disparity of energy intake and its disbursement. This condition is linked with several metabolic disorders representing a strong risk factor for hypertension, heart disease, hyperlipidemia and type 2 diabetes. Lack of physical activity and imbalanced diet is a major road to obesity. Studies suggest that consumption of anthocyanins improve the job of adipocytes and inhibit the obesity. They increase the levels of adiponectin, an important adipocytokines, that is found at decreased levels in obesity [1].

Animal experiments on mice confirmed the obesity preventive effect of corns containing anthocyanins [2]. Similar anti-obesity effects were observed in rats fed with anthocyanins from black soybean and blueberry. Results obtained by these experiments indicate an increase in serum high-density lipoprotein (HDL)-and a decrease in cholesterol triglyceride and cholesterol levels [3].

Tsuda et al. (2005) studied the effect of anthocyanins on gene expression of cultured adipocyte, it has been reported that whole blueberry results increase in obesity while purified blueberry anthocyanins has significant anti-obesity action [4]. The total RNA isolated from the adipocytes was analyzed using GeneChip microarray. A total of 633 or 427 genes were up-regulated, after the treatment of adipocytes with 100 μM of cyanidin-3-glucoside or cyanidin, respectively. Based on the gene expression profile, the up-regulation of hormone-sensitive lipase and enhancement of the lipolytic activity were suggested to be the result of anthocyanin treatment on adipocytes [4]. An excessive adipose tissue (AT) accumulation has metabolic consequences, as adipocyte dysfunction, strongly associated with the development of obesity and diabetes, involving insulin resistance. However, few studies suggest that AT is an important site of *Vaccinium* species actions to ameliorate obesity complications [5]. Anthocyanins were shown to regulate obesity and insulin sensitivity associated with adipocytokine secretion and PPARc activation in adipocytes [6]. High fat diet along with blueberry attenuates insulin resistance and hyperglycemia in mice coincident with reductions in adipocyte death [7]. In another animal study, Prior et al. (2010) concluded that consumption of purified anthocyanins (0.2 mg/ml) in the drinking water (0.49 mg/mouse/day) improved β-cell function and, the rate of fat deposition was decreased, however, blueberry juice was not effective in preventing obesity. Interestingly enough, lower serum leptin concentrations were found in anthocyanins treatments, which retarded the development of obesity [3].

According to Titta et al. (2009) [8], orange juice has anti-obesity effect on fat accumulation but only anthocyanins are not responsible, multiple components present in the orange juice might act synergistically to inhibit fat accumulation. Animal studies using *Vaccinium asheii* cultivars, results in reduction in food intake and, consequently a decrease in body weight gain [9].

8.3 Anthocyanins and Diabetes

Diabetes is a metabolic disease that occurs by a combination of several genetic and life-style patterns, which leads to hyperglycemia and associated symptoms including polydipsia, polyuria and polyphagia (Fig. 8.1). Ultimately, hyperglycemia seriously damages nerves and blood vessels and can lead to irritability and blurred vision. The WHO estimates that number of diabetic patients will be doubled within coming two decades. This exponential rise is mainly due to lack of healthy balanced diet, sedentary life style and increasing prevalence of obesity [10–14].

Diabetes may arise due to persistent stress on pancreas, degeneration of β-cells, decreased insulin secretion, and increased insulin resistance [15–19]. Various scientists have demonstrated anti-diabetic action of anthocyanins, which can be ascribed to several and concurrent effects of anthocyanins, including decreasing glucosuria, Hb A1c and blood glucose, increasing insulin secretion, improving insulin resistance and preventing excessive generation of free radicals [20, 21] however plasma insulin or blood glucose concentrations remain the same in healthy humans [22]. In addition, anthocyanins may contribute to the prevention of type-2 diabetes through its antioxidant activity which may protect β-cells from glucose induced oxidative stress [23]. Sugimoto et al. (2003) reported that boysenberry anthocyanins reinstated or favored to reinstate the biomarkers of oxidative stress to the level of the control group in STZ-induced diabetic rats [24].

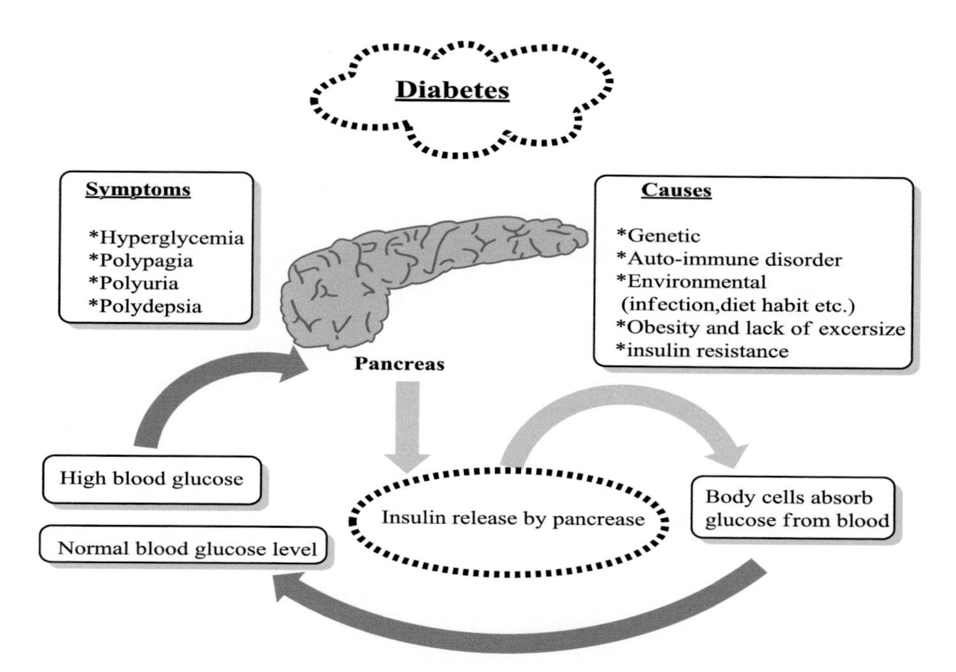

Fig. 8.1 Pathogenesis of diabetes and glucose regulation by pancreas

Stull et al. (2010) suggested that consumption of blueberries increase insulin action or sensitivity effectiveness in increasing insulin action in vivo, in obese and insulin resistant (non-diabetic) human subjects, without changes in adiposity or inflammatory biomarkers [25]. However, more cellular mechanistic studies are needed to clarify the specific cellular pathway involved in improving sensitivity to insulin when blueberries were consumed. Furthermore, hypoglycemic activity has been reported by various scientists as anthocyanins and anthocyanidins induce insulin secretion and the order of affect observed was delphinidin-3-glucoside > cyanidin-3-glucoside > and pelargonidin-3-galactoside [20]. Grace et al. (2009) reported anthocyanins enriched fraction from low bush black berries for hypoglycemic activity in acute mouse model of type 2 diabetes however the fruit extract has no hypoglycemic effect [20]. Among these studies, Tsuda et al. (2003) evaluated that supplementation with cyaniding-3-O-α-D-glucoside rich purple corn color improves hyperglycemia induced by high fat diet in mice [26]. Another study by Vuong et al. (2010) showed that biotransformed blueberry juice decreases hyperglycemia in a diabetic animal model, by partly reversing adiponectin levels thus protects young pre-diabetic mice from developing diabetes and obesity [27].

Tsuda et al. (2003) demonstrated that dietary anthocyanins significantly normalized hypertrophy of the adipocytes in the epididymal white adipose tissue with increase in adiponectin and leptin secretion, thus diminishing hyperglycemia inflicted by the high fat diet in mice (C57BL/6J) [26]. Later in 2005 Tsuda et al. reported that if we treat adipocytes with cyaniding-3-glucoside or cyanidin, the hormone sensitive lipase gene expression elevation occur that results in lipolytic activity and insulin sensitivity [4, 26].

Suzuki et al. (2011) determined that bilberry extracts are capable of inhibiting adipocyte differentiation of 3T3-L1 cells in a dose-dependent manner and to diminish lipid accumulation with concomitant down-regulation of peroxisome proliferator-activated receptor gamma (PPARc). Additionally, all tested single anthocyanins (pelargonidin, cyanidin, delphinidin, peonidin, and malvinidin) also inhibit lipid accumulation in 3T3-L1 cells but delphinidin was most effective in down-regulating PPARc and sterol regulatory element-binding protein 1c (SREBP1c) mRNA levels [5]. Bilberry extracts inhibit adipocyte differentiation via insulin pathway and these effects are mainly due to the presence of anthocyanins. Clinical trials involving blueberry consumption or blueberry extract are recommended to get benefits of obesity prevention and to understand it mechanism [9]. A simplified way of anthocyanins to control diabetes is shown in Fig. 8.2.

8.3.1 Oxidative Stress

There is a vital linkage among oxidative stress, inflammatory response and insulin activity. Excessive generation of free radicals occurs in hyperglycemic and hyperlipidemic state. In order to reduce free radicals and ultimately stress, antioxidants like anthocyanins are required to be consumed [24, 28]. Ischemia-reperfusion-induced

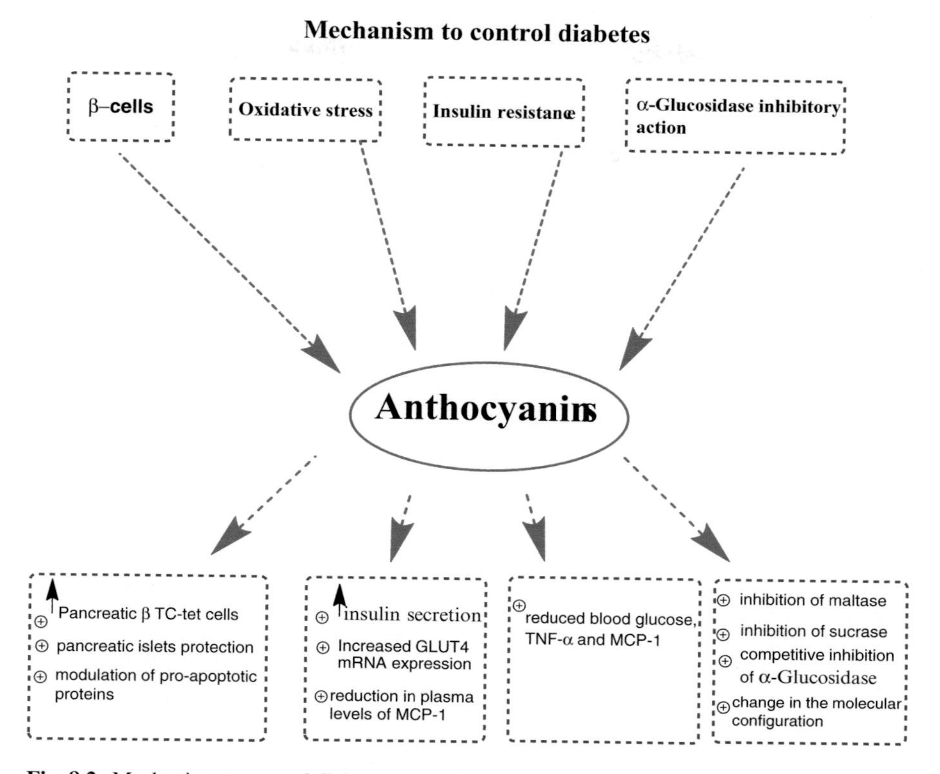

Fig. 8.2 Mechanism to control diabetes via anthocyanins

oxidative stress leads to high concentrations of thiobarbituric acid reactive substances (TBARS) and diminished levels of reduced glutathione [26], the details of anti-diabetic studies in animals which involves reduction of oxidative stress are given in Table 8.1.

Anthocyanins enriched freeze-dried blueberry powder increased plasma antioxidant levels after ingestion of fat diet in healthy men [33]. Similarly juice prepared from freeze-dried strawberry powder lowers lipid peroxidation, leading to decreased plasma levels of oxidized LDL, MDA and 4-hydroxy-noneal in women with metabolic syndrome after consuming two glasses for 4 weeks [22].

8.3.2 *β Cells*

When diabetes type 2 develops pancreatic β cells stop working due to high oxidative stress. But anthocyanins can decrease oxidative stress by their antioxidant action and β cells may be protected [31]. Blueberry (*Vaccinium angustifolium*) fruit extract augmented the proliferation of pancreatic β TC-tet cells exerting a substantial anti-diabetic effect [34] that may be due to anthocyanins [35]. Jayaprakasam et al. (2006)

Table 8.1 Anthocyanins that results reduction in oxidative stress

Anthocyanins	Model	Mechanism	References
Diet containing 0.2 % cyanidin-3-glucoside	Ischemia reperfusion rat model	↓ Serum thiobarbituric acid reactive substances (TBARS)	[26]
		↑ Hepatic reduced glutathione in rats	
Diet containing 0.1 % mulberry anthocyanins	Streptozotocin(STZ)-induced diabetic Wistar rats	↓ Liver oxidized glutathione and tended ↓ TBARS levels	[24]
Diet with 0.5 % of black rice anthocyanins-rich extract	Sprague–Dawley rats	↓ Blood concentrations of TBARS ↓ glutathione	[29]
Diet supplemented with 0.2 % chokeberry fruit extract	Pre-diabetic Wistar rats	Normalize the concentrations of TBARS in the liver, kidneys and lungs and no effect on SOD	[30]
One i.p. injection of pelargonidin at a dose of 3 mg/kg body weight	STZ-induced diabetic Wistar rats	Restoration of exhausted serum levels of SOD and catalase and normalization of the serum levels of MDA	[31]
Anthocyanins from black soybean seed coats	STZ-diabetic rats	Reinstatement of the depleted serum levels of catalyze and SOD enzymes and Normalization of the serum levels of MDA	[32]
Anthocyanins rich ethanolic extract of *Vaccinium arctostaphylos* fruit, Orally	Alloxan-induced diabetic Wistar rats	↑ The concentration of glutathione peroxidase, catalase and superoxide dismutase in red blood cells	[28]

reported that cornelian cherry anthocyanins protected the pancreatic islets of C57BL/6 CB mice and insulin secretion was increased [36]. Treatment with anthocyanins from black soybean seed coats prevented the apoptosis of β cells through modulation of pro-apoptotic proteins caspase 3 and Bax which amplified Bcl2 expression which indicates anti-apoptotic effects [32]. Purified anthocyanins were reported to increase β cell function in C57BL/6J mice [3]. Anthocyanins from Chinese bayberry extract protected pancreatic β INS-1 cells and primary islets from damage induced by H_2O_2 [37].

8.3.3 Insulin Secretion

In type 2 diabetes, insulin secretion may be impaired due to the dysfunction of β cells [38]. Synthetic agents like sulfonyl urea increase insulin secretion but anthocyanins, an important natural bioactives may act as insulin secretagogue without producing such side effect [31, 32]. Delphinidin-3-glucoside in INS-1 832/13 cells led to decreased insulin secretion more than cyaniding-3-glucoside which is more potent than pelargonidin-3-galactoside. Among various anthocyanins evaluated, only pelargonidin showed secretagouge activity. Pelargonidin-3-glucoside and anthocyanins extracted from black soybean increased insulin secretion in STZ-induced-diabetic Sprague–Dawley and Wistar diabetic rats respectively [31, 32]. Extract of *V. arctostaphylos* increases insulin expression in cardiac and pancreatic cells of alloxon-induced-diabetic Wistar rats [28].

8.3.4 Insulin Resistance

Anthocyanins have been reported for anti-inflammatory potential and can decrease the expression of monocyte chemotactic protein-1 (MCP-1), tumor necrosis factor α (TNF-α), interleukin-6 (IL-6), thus improving T2D and insulin resistance [7, 39]. The proposed hypoglycemic activity of anthocyanins can be ascribed partly by the activation of insulin receptors generated [32]. Anthocyanins can modify the expression of adipocytokines, intensify the expression of GLUT4, reduce the expression of RBP4, stimulate AMPK and decrease the oxidative stress to improve insulin sensitivity, thus positively modulating the glycemic control. Table 8.2 indicates improvement of insulin resistance in anthocyanins treated animal models.

8.3.5 α-Glucosidase Inhibitory Action

α-Glucosidases cause hydrolysis of carbohydrates in small intestine, thus inhibiting these enzymes is one therapeutic approach to control diabetes specially after meal [42]. Matsui et al. (2002) reported that anthocyanins extract of purple sweet potato reduces the blood glucose through inhibition of maltase activity in healthy Sprague–Dawley rats after carbohydrate loading (sucrose, maltose and glucose) [43]. Another study by Jurgoński et al. (2008) reported that anthocyanins from chokeberry lowered the high glucose level by inhibiting maltase and sucrase in Wistar rat model of prediabetes obtained through a high-fructose diet and simultaneous single injection of streptozotocin (20 mg/kg) [30]. Anthocyanins cause inhibition of α-glucosidases [43] but the exact mechanism of inhibition is unknown, however it is considered to be competitive, since normal substrates of the enzyme and glucosyl groups linked to anthocyanins are structurally similar [44]. Another possible mechanism may be change in the molecular conformation of the enzyme that polar groups of enzyme active site interact with hydroxyl groups on anthocyanins, leading to a modification in enzyme activity [45]. Another important aspect to be considered is that the use of synthetic inhibitors of α-glucosidases and anthocyanins might have synergistic effect and the α-glucosidases inhibition of anthocyanins is directly related to its stability [46].

Table 8.2 Insulin resistance improvement in various models

Anthocyanins	Model	Insulin resistance improvement mechanism	Reference
High-fat and anthocyanins supplemented diet	C57BL/6J mice	Reduced blood glucose, TNF-α	[26]
Anthocyanins supplemented diet	KK-Ay mice	Reduced blood glucose, TNF-α and MCP-1	[40]
High-fat diet supplemented with tart cherry powder	Zucker rats	Reduced glycemia and insulinemia as well as improved insulin resistance due to a decrease in plasma levels of IL-6 and TNF-α	[39]
High-fat diet supplemented with freeze-dried whole blueberry powder	C57BL/6 mice	Reduced levels of fasting glucose and TNF-α expression in adipose tissue and improved insulin tolerance test (ITT)	[7]
Freeze-dried powders from Different anthocyanins-rich whole berries in hypo- and hyper-lipidemic diets	C57BL/6 J mice	No changes in plasma levels of TNF-α, IL-6 and PAI-1	[41]
High-fat diet and drinking water supplemented with anthocyanins purified from blueberries	C57BL/6 J mice	↓ Plasma levels of MCP-1	[41]
Anthocyanins extracted from black soybean	Sprague–Dawley rats	↑Autophosphorylation of insulin receptors	[32]
Anthocyanins from different plant species	Animal models	↓ Glycemia and ↑expression of the glucose transporter type 4 (GLUT4)	[21, 32, 40]
Extract of *Vaccinium arctostaphylos*	Heartcells of alloxan-induced diabetic Wistar rats	IncreasedGLUT4 mRNA expression	[28]
Billberry extract in diet	DiabeticKK-Ay mice	↓ Plasma glucose and ↑ insulin sensitivity in activation of AMPK in white adipose, skeletal muscle and liver tissues	[21]

8.3.6 Obesity and Diabetes

Due to sedentary life-style and changed dietary patterns especially large intake of refined food has led to a dramatic rise in obesity and diabetes. Obesity and diabetes are inter connected disorders and anthocyanins have been reported for health beneficial effect in association with both for example Tsuda et al. (2003, 2008) reported high fat diet if supplemented with cyanidin-3-glucoside cause reduction in fat

Table 8.3 Anthocyanins simultaneous effects on obesity and diabetes

Anthocyanins	Model	Effect	Reference
High-fat diet (45 % of energy) with anthocyanins extracted from blueberries	C57BL/6 mice	Inhibit weight gain and body fat accumulation	[47]
Whole blueberry powder	C57BL/6 mice	Promote body fat accumulation	[47]
Blueberry juice high fat diet (45 % of energy)	Mice	Reduction in body weight gain and percentage of white adipose tissue	[3]
Supplementing a high-fat diet (60 % of energy) with WBP	C57BL/6 mice	Did not ↓ body weight gain	[7]
Supplementing a high-fat diet (45 % of energy) with 2 % WBP	Zucker fatty rats	Reduced abdominal fat mass and increased the activity of the adipose tissue and skeletal muscle peroxisome proliferator-activated receptor	[48]
Supplementing a high-fat diet (45 % of energy) with 2 % WBP	Zucker lean rats	WBP-induced body weight gain	[48]
Black raspberry anthocyanins (as a juice or in powder form) high-fat diet (60 % of energy)	Mice	Didnot significantly reduce body fat accumulation or body weight gain	[3, 41, 49]
Mulberry water extracts high concentration of anthocyanins		Decrease body weight gain	[50]
Tart cherry powder	Zucker fatty rats	Decrease body weight gain	[39]
Anthocyanins-rich berry, chokeberry fructose rich diet	Mice	Suppress the increase in epididymal white adipose tissue and blood glucose level	[51]

accumulation in C57BL/6J mice [2, 26] and suppression of elevated serum glucose levels induced by high-fat meals. Table 8.3 shows simultaneous effects of anthocyanin in diabetes and obesity.

8.4 Improvement of Eye Vision

The mounting evidence supported by clinical trials indicates anthocyanins as a potential future drug to treat various opthalmological diseases. Bilberry anthocyanins increase night vision by interacting with rhodopsin [52] or phosphodiesterase (PDE) on photo-transduction [53, 54].

Berry anthocyanins improve vision by multiple mechanisms by (i) reducing molecular degeneration and diabetic retinopathy (ii) increasing blood circulation in retina capillaries (iii) improving night-vision by increased production of retinal pigments and by (iv) preventing cataracts, retinitis pigmentosa and glaucoma [55, 56].

Anecdotal evidence in Japan describes Royal Air Force pilots during World War II who experienced enhanced vision in the dark after ingesting blueberry jam indicating that anthocyanin ingestion can increase eye vision [57]. Decreasing the dark-adaptation threshold was noted in a placebo-controlled cross-over double-blind study involving healthy human subjects that were fed blackcurrant anthocyanin concentrate [58]. The effect with the highest dose (50 mg per subject) had a statistically significant effect (P=0.011). Recently, very low (pmol g^{-1}) concentration of blueberry ANs in pig eyes were found after feeding for 4 weeks [59].

Another placebo-controlled double-blind study indicated that oral administration of anthocyanin is vital for the production of visual purple, which helps in conversion of light into electrical signals for brain. Adapto-electro-retinograms of two sets of six subjects were performed before treatment and 1 and 3 h after administration; the subjects consuming bilberries adapted to the light in 6.5 min compared to 9 min for the control group [55, 56]. In another clinical trial of 4 months, 50 patients suffering from sensile cataract were given a mixture of bilberry extract, containing 25 % anthocyanins (@180 mg twice/day) and vitamin E (@100 mg twice/day). This combination stopped the development of cataract in 96 % of the subjects as compared to 76 % in control group [60].

Sundalius (2008) during her PhD research work investigated the ability of blueberry anthocyanins to inhibit the uptake of *N*-retinyl-*N*-retinylidene ethanolamine (A_2E) by retinal pigment epithelial (RPE) cells as quantified by pigment epithelial derived factor (PEDF) levels in RPE cells. The average PEDF results showed that blueberry anthocyanins stimulated and sustained PEDF levels as compared to control RPE cells and RPE cells treated with A_2E thus it was evaluated that blueberry anthocyanins preserved RPE cells from oxidation as indicated by PEDF values [61].

Bilberry anthocyanins have been shown to exhibit neuroprotective effects in ocular tissue and suppressive effects on diabetic retinopathy via a reduction of angiogenesis [62, 63]. Furthermore, several studies have demonstrated that blackcurrant anthocyanins improve visual function; for example, BA inhibit transient myopia, reduce eye fatigue, improve dark adaptation, and increase retinal blood flow in glaucoma [58, 64]. The first three benefits have been shown in humans, and effective daily BA intake was reported as 50 mg [58]. Two studies have indicated the possible mechanism underlying inhibition of myopia. The blackcurrant anthocyanins concentration was higher in ocular tissues (sclera, choroid, ciliary body, retina, iris, and cornea) than in blood in a rat fed a diet containing blackcurrant anthocyanins, and in particular, concentrations in the sclera and choroid were 100-fold higher than the blood blackcurrant anthocyanins concentration [65]. In addition, it was shown that an anthocyanins in the concentration range of 10^{-7} to 10^{-8} M relaxed endothelin-1-induced contraction of bovine ciliary smooth muscle [66], which plays an important role in modulating refraction of the lens through contraction and relaxation, which in turn control accommodation. Thus, it is suggested that anthocyanins stimulate the endothelin-1 receptor to induce production of NO, thereby relaxing ciliary smooth muscle, which in turn flattens the lens and consequently inhibits myopia.

Vaccinium myrtillus extracts containing anthocyanins were reported to improve the night vision, particular diabetic retinopathy, through the influence on the permeability of retinal vessels which also decrease the permeability of the blood-brain barrier to toxic substances [67]. The differences in results and conclusions are probably due to variations in dose and source of anthocyanin, type of subjects and the techniques used to evaluate vision.

8.5 Conclusions

Health is a key priority of people of all age and income group and a vital feature of a prosperous life. Health-concerned consumers of modern society are now moving from curative to preventive health measures. It has led to the generation of dietary recommendation and food composition tables. It is well-known now that plant-based foods and bioactive compounds found in them like anthocyanins help in remaining healthy. Anthocyanins act as preventive and curative agents for various diseases and disorders as presented in this chapter. Works that suggested protective actions of anthocyanins are mainly based on in vitro experiments, a fact that compromises the robustness of these findings. More details are required to establish the real implications of anthocyanins and the mechanism by which they exert their biological effects.

References

1. Arita, Y., Kihara, S., Ouchi, N., Takahashi, M., Maeda, K., Miyagawa, J., et al. (1999). Paradoxical decrease of an adipose-specific protein, adiponectin, in obesity. *Biochemical and Biophysical Research Communications, 257*(79–83), 466.
2. Tsuda, T. (2008). Regulation of adipocyte function by anthocyanins; possibility of preventing the metabolic syndrome. *Journal of Agricultural and Food Chemistry, 56*(3), 642–646.
3. Prior, R. L., Wilkes, S. E., Rogers, T. R., Khanal, R. C., Wu, X., & Howard, L. R. (2010). Purified blueberry anthocyanins and blueberry juice alter development of obesity in mice fed an obesogenic high-fat diet. *Journal of Agricultural and Food Chemistry, 58*(7), 3970–3976.
4. Tsuda, T., Ueno, Y., Kojo, H., Yoshikawa, T., & Osawa, T. (2005). Gene expression profile of isolated rat adipocytes treated with anthocyanins. *Biochimica et Biophysica Acta, 1733*(2), 137–147.
5. Suzuki, R., Tanaka, M., Takanashi, M., Hussain, A., Yuan, B., Toyoda, H., et al. (2011). Anthocyanidins-enriched bilberry extracts inhibit 3T3-L1 adipocyte differentiation via the insulin pathway. *Nutrition and Metabolism, 8*(1), 14.
6. Tsuda, M., Mizokoshi, A., Shigemoto-Mogami, Y., Koizumi, S., & Inoue, K. (2004). Activation of p38 mitogen activated protein kinase in spinal hyperactive microglia contributes to pain hypersensitivity following peripheral nerve injury. *Glia, 45*(1), 89–95.
7. DeFuria, J., Bennett, G., Strissel, K. J., Perfield, J. W., Milbury, P. E., Greenberg, A. S., et al. (2009). Dietary blueberry attenuates whole-body insulin resistance in high fat-fed mice by reducing adipocyte death and its inflammatory sequelae. *Journal of Nutrition, 139*(8), 1510–1516.
8. Titta, L., Trinei, M., Stendardo, M., Berniakovich, I., Petroni, K., Tonelli, C., et al. (2009). Blood orange juice inhibits fat accumulation in mice. *Int J Obesity, 34*(3), 578–588.

9. Norberto, S., Sara, S., Manuela, M., Ana, F., Manuela, P., & Conceição, C. (2013). Blueberry anthocyanins in health promotion: A metabolic overview. *Journal of Functional Foods, 5*(4), 1518–1528.
10. Saad, B., & Said, O. (2011). *Greco-Arab and Islamic herbal medicine: Traditional system, ethics, safety, efficacy, and regulatory issues*. Hoboken: Wiley.
11. Saad, B., Zaid, H., & Said, O. (2012). *Tradition and perspectives of diabetes treatment in Greco-Arab and Islamic medicine.* In Bioactive food as dietary interventions for diabetes: Bioactive foods in chronic disease states (pp. 319–326).
12. Zaid, H., & Saad, B. (2012). *State of the art of diabetes treatment in Greco-Arab and Islamic medicine.* In Bioactive food as dietary interventions for diabetes: Bioactive foods in chronic disease states (p. 327).
13. Saad, B. (2015). *Integrating traditional Greco-Arab and Islamic diet and herbal medicines in research and clinical practice.* In Phytotherapies: Efficacy, safety, and regulation (p. 142).
14. Kadan, S., Saad, B., Sasson, Y., & Zaid, H. (2013). In vitro evaluations of cytotoxicity of eight antidiabetic medicinal plants and their effect on GLUT4 translocation. *Evidence-Based Complementary and Alternative Medicine, 2013.*
15. Sultan, M. T., Butt, M. S., Karim, R., Iqbal, S. Z., Ahmad, S., Zia-Ul-Haq, M., et al. (2014). Effect of *Nigella sativa* fixed and essential oils on antioxidant status, hepatic enzymes, and immunity in streptozotocin induced diabetes mellitus. *BMC Complementary and Alternative Medicine, 14*(1), 193.
16. Bukhari, S. A., Shamshari, W. A., Ur-Rahman, M., Zia-Ul-Haq, M., & Jaafar, H. Z. (2014). Computer aided screening of secreted frizzled-related protein 4 (SFRP4): A potential control for diabetes mellitus. *Molecules, 19*(7), 10129–10136.
17. Khanra, R., Dewanjee, S., Dua, T. K., Sahu, R., Gangopadhyay, M., De Feo, V., et al. (2015). *Abroma augusta* L. (Malvaceae) leaf extract attenuates diabetes induced nephropathy and cardiomyopathy via inhibition of oxidative stress and inflammatory response. *Journal of Translational Medicine, 13*(1), 1–14.
18. Sultan, M. T., Butt, M. S., Karim, R., Zia-Ul-Haq, M., Batool, R., Ahmad, S., et al. (2014). Nigella sativa fixed and essential oil supplementation modulates hyperglycemia and allied complications in streptozotocin-induced diabetes mellitus. *Evidence-Based Complementary and Alternative Medicine, 2014.*
19. Zia-Ul-Haq, M., Ahmad, S., Bukhari, S. A., Amarowicz, R., Ercisli, S., & Jaafar, H. Z. (2014). Compositional studies and biological activities of some mash bean (*Vigna mungo* (L.) Hepper) cultivars commonly consumed in Pakistan. *Biological Research, 47*, 23.
20. Grace, M. H., Ribnicky, D. M., Kuhn, P., Poulev, A., Logendra, S., Yousef, G. G., et al. (2009). Hypoglycemic activity of a novel anthocyanin-rich formulation from lowbush blueberry, *Vaccinium angustifolium* Aiton. *Phytomedicine, 16*(5), 406–415.
21. Takikawa, M., Inoue, S., Horio, F., & Tsuda, T. (2010). Dietary anthocyanin-rich bilberry extract ameliorates hyperglycemia and insulin sensitivity via activation of AMP-activated protein kinase in diabetic mice. *Journal of Nutrition, 140*(3), 527–533.
22. Basu, A., & Penugonda, K. (2009). Pomegranate juice: A heart-healthy fruit juice. *Nutrition Reviews, 67*(1), 49–56.
23. Al-Awwadi, N. A., Araiz, C., Bornet, A., Delbosc, S., Cristol, J. P., Linck, N., et al. (2005). Extracts enriched in different polyphenolic families normalize increased cardiac NADPH oxidase expression while having differential effects on insulin resistance, hypertension, and cardiac hypertrophy in high-fructose-fed rats. *Journal of Agricultural and Food Chemistry, 53*(1), 151–157.
24. Sugimoto, M., Kuo, M. L., Roussel, M. F., & Sherr, C. J. (2003). Nucleolar Arf tumor suppressor inhibits ribosomal RNA processing. *Molecular Cell, 11*(2), 415–424.
25. Stull, A. J., Cash, K. C., Johnson, W. D., Champagne, C. M., & Cefalu, W. T. (2010). Bioactives in blueberries improve insulin sensitivity in obese, insulin-resistant men and women. *Journal of Nutrition, 140*(10), 1764–1768.
26. Tsuda, T., Horio, F., Uchida, K., Aoki, H., & Osawa, T. (2003). Dietary cyanidin 3-O-β-D-glucoside-rich purple corn color prevents obesity and ameliorates hyperglycemia in mice. *Journal of Nutrition, 133*(7), 2125–2130.

27. Vuong, T., Matar, C., Ramassamy, C., & Haddad, P. S. (2010). Biotransformed blueberry juice protects neurons from hydrogen peroxide-induced oxidative stress and mitogen-activated protein kinase pathway alterations. *British Journal of Nutrition, 104*(05), 656–663.
28. Feshani, A. M., Kouhsari, S. M., & Mohammadi, S. (2011). *Vaccinium arctostaphylos*, a common herbal medicine in Iran: Molecular and biochemical study of its antidiabetic effects on alloxan-diabetic Wistar rats. *Journal of Ethnopharmacology, 133*(1), 67–74.
29. Guo, H., Ling, W., Wang, Q., Liu, C., Hu, Y., Xia, M., et al. (2007). Effect of anthocyanin-rich extract from black rice (*Oryza sativa* L. indica) on hyperlipidemia and insulin resistance in fructose-fed rats. *Plant Foods for Human Nutrition, 62*(1), 1–6.
30. Jurgoński, A., Juśkiewicz, J., & Zduńczyk, Z. (2008). Ingestion of black chokeberry fruit extract leads to intestinal and systemic changes in a rat model of prediabetes and hyperlipidemia. *Plant Foods for Human Nutrition, 63*(4), 176–182.
31. Roy, M., Sen, S., & Chakraborti, A. S. (2008). Action of pelargonidin on hyperglycemia and oxidative damage in diabetic rats: Implication for glycation-induced hemoglobin modification. *Life Sciences, 82*(21), 1102–1110.
32. Nizamutdinova, I. T., Kim, Y. M., Chung, J. I., Shin, S. C., Jeong, Y. K., Seo, H. G., et al. (2009). Anthocyanins from black soybean seed coats stimulate wound healing in fibroblasts and keratinocytes and prevent inflammation in endothelial cells. *Food and Chemical Toxicology, 47*(11), 2806–2812.
33. Kay, C. D., & Holub, B. J. (2002). The effect of wild blueberry (*Vaccinium angustifolium*) consumption on postprandial serum antioxidant status in human subjects. *British Journal of Nutrition, 88*(04), 389–397.
34. Martineau, L. C., Couture, A., Spoor, D., Benhaddou-Andaloussi, A., Harris, C., Meddah, B., et al. (2006). Anti-diabetic properties of the Canadian lowbush blueberry *Vaccinium angustifolium* Ait. *Phytomedicine, 13*(9), 612–623.
35. Prior, R. L., Cao, G., Martin, A., Sofic, E., McEwen, J., O'Brien, C., et al. (1998). Antioxidant capacity as influenced by total phenolic and anthocyanin content, maturity, and variety of *Vaccinium* species. *Journal of Agricultural and Food Chemistry, 46*(7), 2686–2693.
36. Jayaprakasam, B., Olson, L. K., Schutzki, R. E., Tai, M. H., & Nair, M. G. (2006). Amelioration of obesity and glucose intolerance in high-fat-fed C57BL/6 mice by anthocyanins and ursolic acid in Cornelian cherry (*Cornus mas*). *Journal of Agricultural and Food Chemistry, 54*(1), 243–248.
37. Zhang, B., Kang, M., Xie, Q., Xu, B., Sun, C., Chen, K., et al. (2010). Anthocyanins from Chinese bayberry extract protect β cells from oxidative stress-mediated injury via HO-1 upregulation. *Journal of Agricultural and Food Chemistry, 59*(2), 537–545.
38. Ahrén, B., Pacini, G., Foley, J. E., & Schweizer, A. (2005). Improved meal-related β-cell function and insulin sensitivity by the dipeptidyl peptidase-IV inhibitor vildagliptin in metformin-treated patients with type 2 diabetes over 1year. *Diabetes Care, 28*(8), 1936–1940.
39. Seymour, E. M., Lewis, S. K., Urcuyo-Llanes, D. E., Tanone, I. I., Kirakosyan, A., Kaufman, P. B., et al. (2009). Regular tart cherry intake alters abdominal adiposity, adipose gene transcription, and inflammation in obesity-prone rats fed a high fat diet. *Journal of Medicinal Food, 12*(5), 935–942.
40. Sasaki, R., Nishimura, N., Hoshino, H., Isa, Y., Kadowaki, M., Ichi, T., et al. (2007). Cyanidin 3-glucoside ameliorates hyperglycemia and insulin sensitivity due to down regulation of retinol binding protein 4 expression in diabetic mice. *Biochemical Pharmacology, 74*(11), 1619–1627.
41. Prior, R. L., Wu, X., Gu, L., Hager, T., Hager, A., Wilkes, S., et al. (2009). Purified berry anthocyanins but not whole berries normalize lipid parameters in mice fed an obesogenic high fat diet. *Molecular Nutrition and Food Research, 53*(11), 1406–1418.
42. Chiasson, J. L., Josse, R. G., Gomis, R., Hanefeld, M., Karasik, A., Laakso, M., et al. (2002). Acarbose for prevention of type 2 diabetes mellitus: The STOP-NIDDM randomised trial. *The Lancet, 359*(9323), 2072–2077.
43. Matsui, T., Ebuchi, S., Kobayashi, M., Fukui, K., Sugita, K., Terahara, N., et al. (2002). Anti-hyperglycemic effect of diacylated anthocyanin derived from *Ipomoea batatas* cultivar

Ayamurasaki can be achieved through the α-glucosidase inhibitory action. *Journal of Agricultural and Food Chemistry, 50*(25), 7244–7248.

44. McDougall, G. J., & Stewart, D. (2005). The inhibitory effects of berry polyphenols on digestive enzymes. *Biofactors, 23*(4), 189–195.
45. Adisakwattana, S., Charoenlertkul, P., & Yibchok-anun, S. (2009). α-Glucosidase inhibitory activity of cyanidin-3-galactoside and synergistic effect with acarbose. *Journal of Enzyme Inhibition and Medicinal Chemistry, 24*(1), 65–69.
46. Sancho, R. A. S., & Pastore, G. M. (2012). Evaluation of the effects of anthocyanins in type 2 diabetes. *Food Research International, 46*(1), 378–386.
47. Prior, R. L., Wu, X., Gu, L., Hager, T. J., Hager, A., & Howard, L. R. (2008). Whole berries versus berry anthocyanins: Interactions with dietary fat levels in the C57BL/6 J mouse model of obesity. *Journal of Agricultural and Food Chemistry, 56*(3), 647–653.
48. Seymour, E. M., Tanone, I. I., Urcuyo-Llanes, D. E., Lewis, S. K., Kirakosyan, A., Kondoleon, M. G., et al. (2011). Blueberry intake alters skeletal muscle and adipose tissue peroxisome proliferator-activated receptor activity and reduces insulin resistance in obese rats. *Journal of Medicinal Food, 14*(12), 1511–1518.
49. Kaume, L., William, C. G., Cindi, B., Luke, R. H., & Latha, D. (2012). Cyanidin 3-O-β-D-glucoside-rich blackberries modulate hepatic gene expression, and anti-obesity effects in ovariectomized rats. *Journal of Functional Foods, 4*(2), 480–488.
50. Peng, C.-H., Liu, L. K., Chuang, C. M., Chyau, C. C., Huang, C. N., & Wang, C. J. (2011). Mulberry water extracts possess an anti-obesity effect and ability to inhibit hepatic lipogenesis and promote lipolysis. *Journal of Agricultural and Food Chemistry, 59*(6), 2663–2671.
51. Qin, B., & Anderson, R. A. (2012). An extract of chokeberry attenuates weight gain and modulates insulin, adipogenic and inflammatory signalling pathways in epididymal adipose tissue of rats fed a fructose-rich diet. *British Journal of Nutrition, 108*(04), 581–587.
52. Bastide, P., Rouher, F., & Tronche, P. (1967). Rhodopsin and anthocyanosides. Apropos of various experimental facts. *Bulletin des Sociétés d'Ophtalmologie de France, 68*(9), 801–807.
53. Ferretti, C., Magistretti, M. J., Robotti, A. P., & Genazzani, G. E. (1988). *Vaccinium myrtillus* anthocyanosides are inhibitors of cAMP and cGMP phosphodiesterases. *Pharmacological Research Communications, 20*, 150.
54. Virmaux, N., Bizec, J. C., Nullans, G., Ehret, S., & Mandel, P. (1990). Modulation of rod cyclic GMP-phosphodiesterase activity by anthocyanidin derivatives. *Biochemical Society Transactions, 18*(4), 686–687.
55. Camire, M. E. (2000). Bilberries and blueberries as functional foods and nutraceuticals. In G. Mazza & B. D. Oomah (Eds.), *Functional foods: Herbs, botanicals and teas* (pp. 289–319). Lancaster: Technomic.
56. Camire, M. (2000). In G. Mazza & B. D. Oomah (Eds.), *Functional foods: Herbs, botanicals and teas*. Lancaster: Technomic.
57. Canter, P. H., & Ernst, E. (2004). Anthocyanosides of *Vaccinium myrtillus* (Bilberry) for night vision-a systematic review of placebo-controlled trials. *Survey of Ophthalmology, 49*(1), 38–50.
58. Nakaishi, H., Matsumoto, H., Tominaga, S., & Hirayama, M. (2000). Effects of black currant anthocyanoside intake on dark adaptation and VDT work-induced transient refractive alteration in healthy humans. *Alternative Medicine Review, 5*(6), 553–562.
59. Kalt, W., Blumberg, J. B., McDonald, J. E., Vinqvist-Tymchuk, M. R., Fillmore, S. A., Graf, B. A., et al. (2008). Identification of anthocyanins in the liver, eye, and brain of blueberry-fed pigs. *Journal of Agricultural and Food Chemistry, 56*(3), 705–712.
60. Head, K. (2001). Natural therapies for ocular disorders part two: Cataracts and glaucoma. *Alternative Medicine Review, 6*(2), 141–166.
61. Sundalius, N. M. (2008). *Examination of blueberry anthocyanins in prevention of age-related macular degeneration through retinal pigment epithelial cell culture study*. Faculty of the Louisiana State University and Agricultural and Mechanical College in partial fulfillment of

the requirements for the degree of Master of Science in The Department of Food Science by Naomi Marie Sundalius BS, Michigan State University.
62. Matsunaga, N., Tsuruma, K., Shimazawa, M., Yokota, S., & Hara, H. (2010). Inhibitory actions of bilberry anthocyanidins on angiogenesis. *Phytotherapy Research, 24*(S1), S42–S47.
63. Matsunaga, N., Chikaraishi, Y., Shimazawa, M., Yokota, S., & Hara, H. (2010). *Vaccinium myrtillus* (bilberry) extracts reduce angiogenesis in vitro and in vivo. *Evidence-Based Complementary and Alternative Medicine, 7*(1), 47–56.
64. Iida, H., Nakamura, Y., Matsumoto, H., Takeuchi, Y., Harano, S., Ishihara, M., et al. (2010). Effect of black-currant extract on negative lens-induced ocular growth in chicks. *Ophthalmic Research, 44*(4), 242–250.
65. Matsumoto, H., Nakamura, Y., Iida, H., Ito, K., & Ohguro, H. (2006). Comparative assessment of distribution of blackcurrant anthocyanins in rabbit and rat ocular tissues. *Experimental Eye Research, 83*(2), 348–356.
66. Matsumoto, H., Kamm, K. E., Stull, J. T., & Azuma, H. (2005). Delphinidin-3-rutinoside relaxes the bovine ciliary smooth muscle through activation of ET B receptor and NO/cGMP pathway. *Experimental Eye Research, 80*(3), 313–322.
67. Igarashi, Y., Chiba, H., Utsumi, H., Miyajima, H., Ishizaki, T., Gotoh, T., et al. (2000). Expression of receptors for glial cell line-derived neurotrophic factor (GDNF) and neurturin in the inner blood-retinal barrier of rats. *Cell Structure and Function, 25*(4), 237–241.

Chapter 9
Anthocyanins Effects on Carcinogenesis, Immune System and the Central Nervous System

9.1 Introduction

As mentioned in Chap. 1, anthocyanins are phytochemicals that are not required for the immediate survival of the plant but which are synthesized to increase the fitness of the plants to survive by allowing them to interact with their environment, including pathogens and herbivorous and symbiotic insects. In many cases, the effects of these secondary metabolites on the human immune system and central nervous system might be linked either to their ecological roles in the life of the plants or to molecular and biochemical similarities in the biology of plants and higher animals. The health professionals now recognize that in addition to the macromolecules, i.e. carbohydrates, lipids and proteins, there is dearth of phytochemicals like anthocyanins and carotenoids which have major heath promoting effects. Age-related disorders like hypertension, diabetes, cardiovascular diseases, Alzheimer's disease, cataracts, neurodegenerative problems and macular degeneration, and improvement of vision and brain functions may be prevented by taking ample amount of these bioactive constituents.

9.2 Anti-Inflammatory Activity

Inflammation is a multifaceted biological response of vascular tissue to stimulants, irritants or injuries and is linked with instigation and progression of various chronic diseases like cardiovascular diseases, Alzheimer's, disease, diabetes mellitus (type 2) and various types of cancers [1–9]. Therefore, anti-inflammatory agents also can act as anticancer agents. Anti-inflammatory properties of anthocyanins, anthocyadin (purified) and their concerned sources were reported by various scientists. For example, Cy-aglycone was found to exhibits higher anti-inflammatory effects than aspirin in COX-assays [10]. Purified anthocyanin fractions from bilberries, blueberries,

M. Riaz et al., *Anthocyanins and Human Health*, SpringerBriefs in Food, Health, and Nutrition, DOI 10.1007/978-3-319-26456-1_9

blackberries, cranberries, sweet cherries, raspberries, elderberries, strawberries and tart cherries exhibit anti-inflammatory properties as assessed by COX-1 and COX-2 inhibitory assays and it may be due to cyanidin glycosides [11]. Strawberry, blackberry, and raspberry showed the highest anti-inflammatory activity, comparable to that of ibuprofen and naproxen at 10 μM concentrations. In an in vivo study, the therapeutic efficacy of blackberry anthocyanins (Cyanidin-3-glucoside accounts for 80 %) was investigated in rats with carrageenan-induced lung inflammation [12]. Anthocyanins effectively reduced all parameters of inflammation dose-dependently.

Anthocyanidins possess structure dependent anti-inflammatory properties. The action mechanism of this effect is believed to be mediated through inhibiting COX-2 in lipo-poly-saccharide (LPS)-activated cells (RAW-264) or inhibiting inducible nitric acid synthase (iNOS) and mRNA expression in LPS-activated murine-J774 macrophages [13]. Cyclooxygenase-2 is involved in many inflammatory actions. Delphinidinis are the strongest inhibitor of COX-2 expression at mRNA and protein levels and exerts inhibitory effect on degradation of nuclear translocation of p65 and IκB-α [14].

Anthocyanins or anthocyanin-containing extract inhibit pro-inflammatory cytokines in vitro by suppressing NF-κB through down-regulation of mitogen-activated protein kinase (MAPK) pathways [15]. Some authors reported that these actions are due to antioxidant effect but some authors reported anthocyanins can exhibit substantial anti-inflammatory property through signaling pathways responsible for anti-inflammatory action without affecting the in vivo anti-oxidative status or its signaling pathways [16, 17]. Wang and his colleagues [17] found that cyanidin-3-glucoside inhibit COX-2 expression and iNOS by induction of liver-X-receptor-α (LRX-α) activation in THP-1 macrophages. The signaling pathways of the nuclear receptors are activated by anthocyanins effectively, for example LRX α and PPAR γ (peroxisome proliferator-activated receptor γ) that antagonizes in vitro inflammatory gene expression [14]. In carrageenan-induce inflammation in rat lungs, blueberry anthocyanins reduced all inflammation parameters dose-dependently [12]. Anthocyanins obtained from red wine inhibited TNF-α-induced inflammation by modulating the endothelial monocyte chemo-attractant protein-1 [18].

9.3 Anthocyanins and Cancer

Cancer is a principal reason of causality and mortality globally. Numerical figures specify that cancer affects more than 1/3rd of world population and more than 20 % death occur due to this malady. In both benign and malignant cancers, cells divide abnormally by escaping the regulatory control of cell resulting in invasion of other body tissues. Till now, more than 100 various type of cancer have been discovered which have been classified depending upon affected tissue or organ like blood cancer, colon cancer and breast cancer etc. Cancer results from DNA abnormalities Inflammation, unhealthy diet and stress-induced oxidative damages are one of major reasons of cancer. Diet and phytochemicals have been used since dawn of civilization as therapeutic strategy to cure various diseases including cancer. Bioactive constituents present in food commodities like grains, fruits, vegetables, nuts, legumes and herbs encompass effective protection against various types of cancers. Therefore dietary chemo-prevention is now focus of attention globally [19–21].

Anthocyanins, although discovered in 1918, their anticancer effects have been studied and reported only recently. Table 9.1 lists anticancer effects of various anthocyanins. Modulation of carcinogenesis occurs through variety of biological functions of anthocyanins. Anthocyanins are believed to exert anti-cancer effects by multiple mechanisms like by (i) arresting cell cycle by arresting the G1/G0 and G2/M phase (ii) inducing apoptosis and anti-angiogenesis (iii) inhibiting oxidative DNA damage (iv) inducing phase II enzymes for detoxification and (v) by inhibiting COX-2 enzymes. The chemopreventive action of anthocyanins is due to their antioxidant property.

Anthocyadins are more effective in inhibition of cell proliferation than anthocyanins [40]. The latter are more potent anti-cancer agents than other flavonoids [23]. In another study by Kamei and his colleagues, anthocyanins from red wine showed higher anticancer activity than other flavonoids of red or white wine [24]. Anthocyanins fraction from 4 cultivars of muscadine grape exhibited higher antiproliferative activity than crude extract or phenolic acid fraction [28]. Antiproliferative effects of anthocyanins on the colon cancer cells are structure dependant [27]. The proposed mechanism of anticancer effects of anthocyanins is shown in Fig. 9.1.

The work of Jing et al. [27] advocated the additive effect of anthocyanins and other phenolics in combinational usage in antiproliferative studies. Various animal studies showed that anthocyanins have chemopreventive effects in gastrointestinal tract like oral cavity, the esophagus [41], and the colon [30]. This chemopreventive effect may be due to anthocyanins contact directly with the epithelial layer [42] while non-gastrointestinal organs required anthocyanins availability through blood delivery. It was confirmed by strawberry anthocyanins failure to inhibit 4-(methylnitrosamino)-1-(3-pyridyl)-1-butanone-and benzo[a]pyrene-induced lung cancer in a mice model [43].

Anti-cancer activity of anthocyanins may be attributed to the additive effect of multiple mechanisms [30]. Possible mechanisms that have been suggested are given in Table 9.2. Various studies showed that anthocyanins can inhibit the growth of different cancer cells and embryonic fibroblasts, indicating their potential as chemopreventive agents in the form of cheap and safe anticancer dietary supplements [50]. Anthocyanins exert antiproliferative activity against human cancer cells derived from malignant tissues [40]. Anthocyanins were also potent and selective proliferation-inhibitor of human promyelocytic leukemia cells [51]. The inhibitory effect on cancer cells of berry extracts deepens upon the contents and not the composition of anthocyanins [52].

Antiangiogenic effect, increased apoptosis and decreased proliferation were observed in patients suffering from colon cancer administered black raspberry powder daily for several weeks [16]. In another clinical study, 25 colorectal cancer patients were given standardized anthocyanin bilberries extract (mirtocyan) administered daily for 1 week, decrease in proliferation index, and increase in apop-

Table 9.1 Anticancer activities reported from different anthocyanins

Anthocyanins source	Model	Effect	Reference
Anthocyanin-rich extract from chokeberry	Colon cancer HT-29 cells	Induce cell cycle block at G1/G0 and G2/M phases but not in NCW460 normal colonic cells	[22]
Anthocyanins extracted from flower petals	Intestinal carcinoma derived HCT-15 cell line	Cell growth inhibition	[23]
Anthocyanins from red wine or white wine	Humangastric cancer HCT-15 cell line and AGS cell line	Cell growth inhibition	[24]
Commercial bilberry, chokeberry, and grape extracts	HT-29 cell line	Cell growth inhibition	[25]
Four anthocyanins isolated from strawberry	Human oral (CAL-27, KB), colon (HT29, HCT-116), and prostate (LNCaP, DU145) cancer cells	Reduction in cell viability	[26]
Anthocyanins rich extract of purple corn, purple carrot, and red radishes	HT-29 cell line	Cell growth inhibition	[27]
Anthocyanins fraction from muscadine grapes (4 cultivars)	Human colon cancer derived cell liens, HT-29 and Caco-2	Cell growth inhibition	[28]
Cranberry extract versus its flavonol glycosides (gly), anthocyanins, proanthocyanidins, and organic acids fractions	Human oral (KB, CAL27), colon (HT-29, HCT116, SW480, SW620), and prostate (RWPE-1, RWPE-2, 22Rv1) cancer cell lines	Anthocyanins and the proanthocyanidin fraction exhibited substantial inhibitory effect except SW480 cell lines, but combination of both exhibited inhibition of all comparatively	[29]
Dietary anthocyanins	Colon carcinogen azoxymethane (AOM)-induced rat colon cancer model	Inhibitory effect on cell proliferartion	[30]
Anthocyanins rich extracts of grapes, bilberries and chokeberries	Human malignant HT-29 colon cancer cells	Cell growth inhibition	[25]
Anthocyanins rich extracts of grapes, bilberries and chokeberries	Low and high tumorigenic colon cancer cell lines, LoVo/Adr and LoVo	Cell growth inhibition	[31]

(continued)

Table 9.1 (continued)

Anthocyanins source	Model	Effect	Reference
Anthocyanins from tart cherries	Human colon cancer cells HT29 and HCT-116	Reduced proliferation	[32]
Anthocyanins extract from *Vaccinium uliginosum*	Human colorectal cancer cells DLD-1 and COLO205	Growth suppression in a dose-dependent manner	[33]
Red grape pomace extract (oenocyanin)	Adenoma development in the ApcMin mouse	Suppress adenoma cell proliferation and ↓-regulation of expression of the PI3 pathway component Akt,	[34]
Vaccinium myrtillus,Vaccinium vitis-idaea and *Rubus chamaemorus*, rich in anthocyanins	Min/1 mice	Chemopreventive as significant ↓ in the number of intestinal tumors	[35]
Grape juice comprising 15 different anthocyanins	Rats mammary tumors	Incidence, multiplicity and final mass reduction	[36]
Lyophilized black raspberries	NMBA (Nitrosomethylbenzylamine)-induced esophageal tumors	Prevention of tumor development	[37]
Anthocyanins containing pomegranate extract	DMBA (7,12-dimethylbenzanthracene)-induced skin tumors in CD-1 mice.	Deferred the onset and reduced the incidence of	[16]
Black rice anthocyanins	Human colon cancer cells HT-29 and HCT-116	Reduced the expression of MMP-2 and MMP-9 in cancer cells	[38]
Georgian Grown Blueberries	HT-29 colon cancer cells	Induce apoptosis decrease GST activities	[39]

totic index was observed [53]. No substantial association was observed between anthocyanidins intake and risk for pharyngeal or oral cancer [54]. Similarly no protective effect has been reported against the development of prostate cancer [55]. To get optimum results the anthocyanins containing berry are consumed before, during and after chemotherapy of cancer may maximize chemopreventive effectiveness in humans. The difference in response against various tumors may be due to its bioavailability to the site [56]. Studies are required to evaluate different anthocyanins in combination with various chemotherapeutic agents to maximize rational treatment of cancer. Anthocyanins from highbush blueberry and mulberry fruits (*V. angustifolium*) stopped the proliferation activities of cells [57]. Delphinidin inhibited invasion of human fibrosarcoma cells by down regulation of gene expression of MMP-2/9 [58]. Delphinin is more potent antioxidant due to highest number of OH groups on B ring.

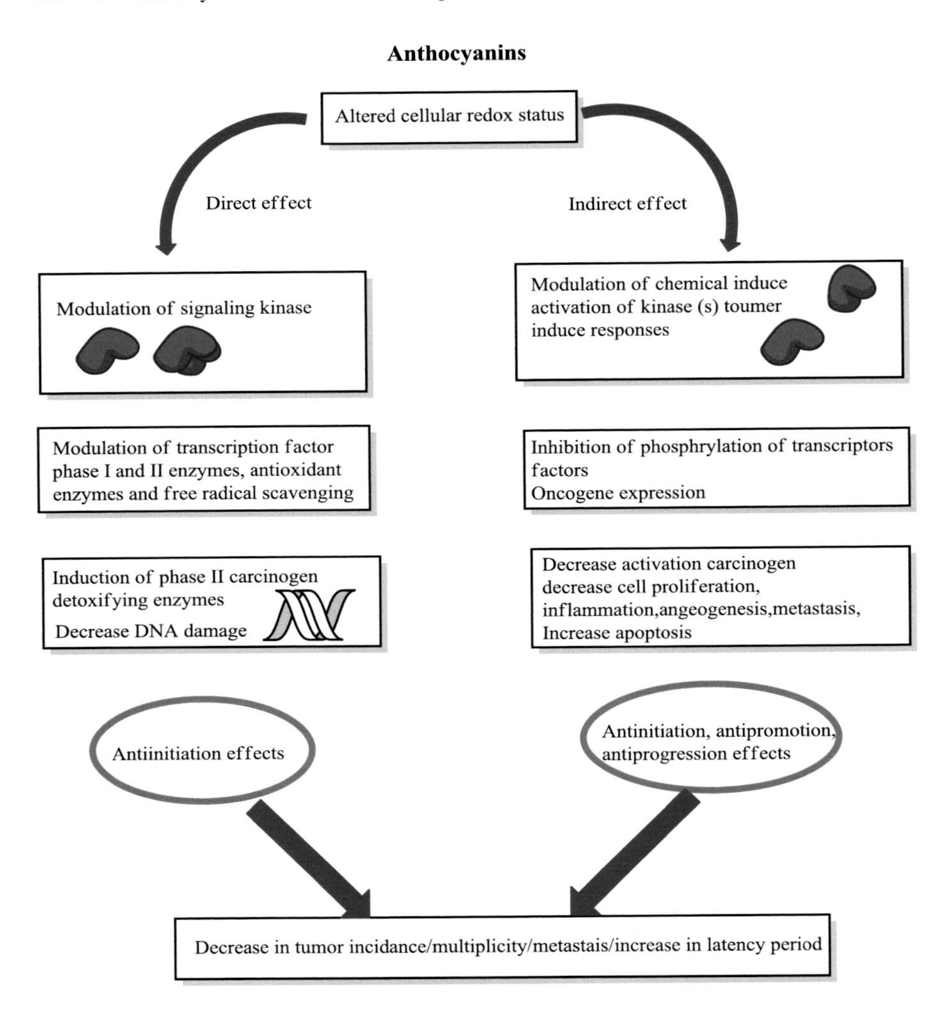

Fig. 9.1 Chemopreventive effect of anthocyanins with mechanism

Table 9.2 Mechanisms of anticancer activity of anthocyanins

Mechanism	Reference
Antimutagenic activity	[44]
Inhibition of oxidative DNA damage	[45]
Induction of phase-II enzymes for detoxification and activation of carcinogen	[46]
Cellcycle arrest	[47]
Induction of apoptosis; inhibition of COX-2 enzymes	[28]
Antiangiogenesis e.g. inhibition of H_2O_2 and tumor necrosis factor α (TNF-α)-induced VEFG-expression, inhibition of VEGF and VEGF-receptor expression	[48]
Increased glutathione S transferase (GST) activity	[49]

9.4 Anthocyanins, Alzheimer Diseases and Brain Function

Anthocyanins prevent Alzheimer's diseases and neuro-degeneration by increasing neuronal signaling in brain, mediating memory function and improving glucose utilization [59]. Anthocyanins have very beneficial effects on memory, cognition and motor function because of delaying the deterioration of neural function [59, 60]. Goyarzu et al. (2004) found in Fisher rats, that blueberry intake suppressed an aging-associated increase in nuclear factor-kappa B levels (NF-κB) [61] while upregulated TNF-α, IL-1β and NF-κB expression in the hippocampus [62]. Williams et al. [63] (2008) demonstrated that blueberry intake induced activation of cyclic AMP-response element-bonding protein and an increase in the level of brain-derived neurotrophic factor. Improvement of cerebrovascular blood flow may also be implicated in the favorable effects of berries on brain function [64]. Ameloriative effects of anthocyanins in neurodegenerative processes of Alzheimer's and Parkinson's disease is chiefly due to their ability to decrease inflammation and oxidative stress in brain. It was observed that berries enhance release of dopamine in human brain leading to improved communication between brain cells. More consumption of anthocyanin-rich diets like vegetables and fruits and can prevent or reverse age-related changes in brain and behaviour [65]. Berries not only stop but also reverse oxidative-stress-induced and age-related deterioration in neural function increasing dopamine release thus improving the ability of neurons to increase intracellular communication [66, 67]. Bilberry-fed animals exhibited improved memory and control of sensory inputs than control animals [66] while rats given lyophilized berries indicated improved working memory and increased short-term memory [68]. In a brain injury mice-model with middle cerebral artery occlusion showed neuroprotective effects when fed mulberry fruit extract containing cyanidin-3-glucoside [69] (Table 9.3).

9.5 Miscellaneous Activities

Anthocyanins can express their antimicrobial activity by causing morphological damages to bacterial cells or by destroying the structural integrity of the wall, membrane, intracellular matrix or by cell deformation [71, 72]. Four anthocyanins namely cyanidin chloride, delphinidin chloride, pelargonidin chloride and cyanidin-3-glucoside possess strong anti-bacterial activity against *E. coli* (gram −ve) strain CM-871 (DNA repair-deficient strain) but did not inhibit normal *Escherichia coli* and beneficial probiotic bacteria (gram +ve). That shows the action association with DNA. In another antibacterial study, anthocyanin exhibited more potential in decreasing the viability of *Salmonella enterica* var. *typhimurium* than other berries phenolic fractions [73]. The effect was due to anthocyanins induction of lipopolysaccharide molecules release from the outer membrane of the gram-negative bacteria.

Table 9.3 Mechanisms of action of anthocyanins [70]

Mechanism	Anthocyanins
Oxidative stress	Scavenges superoxide radicals
	Scavenges hydroxyl radicals
	Increases antioxidant enzyme activity
	Lowers ROS
	Decreases NO production
Cardiomyocytes and the endothelium	Improves endothelial dysfunction
	Mediates vasodilation
Anti-inflammatory effects	Decreases P-selectin, MCP-1, TNF-α and IL-10 expression
	Decreases VEGF and ICAM-1 expression on endothelial cells
	Decreases VLA-4, CD40 and CD36 expression on monocytes
Metabolic effects	Decreases LDL-cholesterol
	Increases HDL-cholesterol
	Inhibits LDL oxidation
	Increases faecal expression of acidic and neutral sterols
	Alters sugar and lipid metabolism
	Improves insulin resistance
	Decreases NFκB levels
	Decreases iNOS and COX_2 expression
Anti-ischemic and cardioprotective effects	Decreases the magnitude of ischemia–reperfusion injury

ROS reactive oxygen species, *NO* nitric oxide, *MCP* monocyte chemotactic protein, *TNF* tumour necrosis factor, *IL* interleukin, *VEGF* vascular endothelial growth factor, *ICAM* intracellular adhesion molecule, *VLA* very late antigen, *CD* cluster of differentiation, *LDL* low density lipoprotein, *HDL* high density lipoprotein, *NFκB* nuclear factor, *iNOS* inducible nitric oxide synthase, *COX* cyclooxygenase

Blueberry and billberry extracts showed growth-inhibitory effects on gram negative bacteria (*Escherchia coli*, *Enterococcus faecali*, *Pseudomonas aeuruginosa*, *Citrobacter freundi* and *Salmonella enterica* ser. *Typhimurium*) and gram positive bacteria (*Staphylococusaureus*, *Listeria monocytogenes*, *Enterococcus faecalis* and *Bacillus subtilis*). Of all bacteria, *E. faecalis* and *C. fruendii* were the most sensitive and *E. coli* showed the highest resistance [74]. American cranberry and European cranberry showed similar anti-bacterial effects [75, 76]. Cranberries decreased growth of *E. coli* below detectable limits when treated with anthocyanins at native pH. Anthocyanins exhibited low antimicrobial activity at neutral pH, probably due to their instability [71]. Cranberry extract showed no effect on yeast. Blackcurrent concentrates arrested the growth of *E. faecium* and *S. aureus* strains while minor effects were observed against *E. coli* [77]. In another comparative study, microbial strains exhibited different susceptibility to berry extracts; the order of antimicrobial effects observed was cloudberries > raspberries > strawberries [73]. In another investigation, berry extracts showed bactericidal effects, inhibiting growth of *Helicobacter pylori*. All berries extracts showed >70 % inhibition at 1 % concentration with bilberry, blueberry and cranberry, extract showing >90 % inhibition [78].

Anthocyanins of berry extracts (bilberry, blueberry, strawberry, cranberry, raspberry) have shown protective effects against liver damage [79], ulcer [80] and UV-radiation-induced damages especially cyaniding acts as skin-protective agent [81].

Several enzymes in vitro may be inhibited by anthocyanins [82] like aldoreductase in the lens of rats can be inhibited by both pelargonidin and delphinidin [83]. Anthocyanin-3-monoglycosides (namely delphinidin-, petunidin-and malvidin-) extracted from grapes inhibited hexokinase, malate dehydrogenase and glycerol dehydrogenase while increased the activity of glutamic acid decarboxylase and α-glucan-phosphorylase [84].

Tobacco bud worm larva growth was reported to be inhibited by cyaniding-3-glucoside, which may be a useful biological controlling agent [85].

9.6 Pharmaceutical Products

Anthocyanin containing diets have been used since long as therapeutic remedy against various diseases in different parts of the world. For example crude extracts of *Vaccinium myrtillus* administered intramuscularly and intravenously or ingested orally can decrease the fragility and permeability of capillaries of blood system in humans. A health-promoting product, OptiBerry is made form the synergistic combination of 6 selected extracts of wild blueberry and bilberry (*Vaccinium myrtillus* and *Vaccinium corymbosum*), elderberries, strawberries, cranberries and raspberries. This product exhibits very high anti-angiogenic, anticarcinogenic, antioxidant, antibacterial, and anti-atherosclerotic properties [86]. OptiBerry shows very low cytotoxicity and can be used as a dietary supplement and safe food. Anthocyanin powder of red cabbage, grape and tomatoes is used as natural food additives in soft drinks and confectionary items. Anthocyanins of red cabbage are stable over a broader pH range and therefore used as substitutes to synthetic blue colorings for foods with neutral pH [87]. Anthocyanins of *Hibiscus* are utilized in herbal formulations as well as in soft drinks [88].

9.7 Conclusions

With better understanding of cancer subtypes and their risk factors, human clinical trials should be designed to obtain maximum benefits from dietary anthocyanins. Research in epigenetics and personalized medicines should be carried out to decipher specific anthocyanins compounds and determine dosage that may modulate risk of various age-related diseases in humans. Taking into account the efficacy and pharmacological effects of anthocyanins in humans, it is believed that combining them with other antioxidants like ascorbic acid will be a useful approach for prevention of various diseases. As they target different signaling pathways providing a complementary and synergistic protective effects against chronic diseases.

References

1. Ahmed, S., Gul, S., Zia-Ul-Haq, M., Riaz, M., & Moga, M. (2015). Anti-inflammatory effects of cyclooxygenase-2 inhibitors in rabbits. *Pakistan Journal of Zoology, 47*(1), 209–216.
2. Ahmed, S., Gul, S., Gul, H., Zia-Ul-Haq, M., Ercisli, S., & Jaafar, H. Z. E. (2014). Scientific basis for the use of *Cinnamonum tamala* in cardiovascular and inflammatory diseases. *Experimental and Clinical Cardiology, 20*, 784–800.
3. Zia-Ul-Haq, M., Khan, B. A., Landa, P., Kutil, Z., Ahmed, S., Qayum, M., et al. (2011). Platelet aggregation and anti-inflammatory effects of garden pea, Desi chickpea and Kabuli chickpea. *Acta Poloniae Pharmaceutica, 69*(4), 707–711.
4. Zia-Ul-Haq, M., Landa, P., Kutil, Z., Qayum, M., & Ahmad, S. (2013). Evaluation of anti-inflammatory activity of selected legumes from Pakistan: In vitro inhibition of Cyclooxygenase-2. *Pakistan Journal of Pharmaceutical Sciences, 26*(1), 185–187.
5. Ahmed, S., Gul, S., Zia-Ul-Haq, M., & Stanković, M. S. (2014). Pharmacological basis of the use of *Acorus calamus* L. in inflammatory diseases and underlying signal transduction pathways. *Latin American and Caribbean Bulletin of Medicinal and Aromatic Plant, 13*(1), 38–46.
6. Ahmed, S., Gul, S., Idris, F., Khan, A. H., Zia-Ul-Haq, M., Jaafar, H. Z. E., et al. (2014). Anti-inflammatory and antiplatelet activities of plasma are conserved across twelve Mammalian species. *Molecules, 19*(8), 11385–11394.
7. Gul, S., Ahmed, S., Kifli, N., Uddin, Q. T., Tahir, N. B., Hussain, A., et al. (2014). Multiple pathways are responsible for anti-inflammatory and cardiovascular activities of *Hordeum vulgare* L. *Journal of Translational Medicine, 12*(1), 1–8.
8. Khanra, R., Dewanjee, S., Dua, T. K., Sahu, R., Gangopadhyay, M., De Feo, V., et al. (2015). *Abroma augusta* L. (*Malvaceae*) leaf extract attenuates diabetes induced nephropathy and cardiomyopathy via inhibition of oxidative stress and inflammatory response. *Journal of Translational Medicine, 13*(1), 1–14.
9. Najm-ur-Rahman, M. R., Khan, A., Zia-Ul-Haq, M., & Dima, L. (2015) Mechanism of anti-inflammatory and anti-nociceptive actions of *Acacia modesta* in animal models. *Pakistan Journal of Zoology, 47*(6), 1723–1730.
10. Wang, H., Nair, M. G., Strasburg, G. M., Chang, Y. C., Booren, A. M., Gray, J. I., et al. (1999). Antioxidant and antiinflammatory activities of anthocyanins and their aglycon, cyanidin, from tart cherries. *Journal of Natural Products, 62*(2), 294–296.
11. Seeram, N. P., Momin, R. A., Nair, M. G., & Bourquin, L. D. (2001). Cyclooxygenase inhibitory and antioxidant cyanidin glycosides in cherries and berries. *Phytomedicine, 8*(5), 362–369.
12. Rossi, A., Serraino, I., Dugo, P., Di, P. R., Mondello, L., Genovese, T., et al. (2003). Protective effects of anthocyanins from blackberry in a rat model of acute lung inflammation. *Free Radical Research, 37*(8), 891–900.
13. Hämäläinen, J. A. (2007). *Processing of sound rise time in children and adults with and without reading problems*. Jyväskylä: University of Jyväskylä.
14. Miguel, M. (2011). *Anthocyanins:* Antioxidant and/or anti-inflammatory activities. *Journal of Applied Pharmaceutical Science, 1*(6), 7–15.
15. Pergola, G., Maldera, S., Tartagni, M., Pannacciulli, N., Loverro, G., & Giorgino, R. (2006). Inhibitory effect of obesity on gonadotropin, estradiol, and inhibin B levels in fertile women. *Obesity, 14*(11), 1954–1960.
16. Wang, Q., Tang, X. N., & Yenari, M. A. (2007). The inflammatory response in stroke. *Journal of Neuroimmunology, 184*(1), 53–68.

17. Wang, X., Jiang, Y., Yu-Wen, W., Mou-Tuan, H., Chi-Tang, H., & Qingrong, H. (2008). Enhancing anti-inflammation activity of curcumin through O/W nanoemulsions. *Food Chemistry, 108*(2), 419–424.
18. Garcia-Alonso, M., Minihane, A. M., Rimbach, G., Rivas-Gonzalo, J. C., & de Pascual-Teresa, S. (2009). Red wine anthocyanins are rapidly absorbed in humans and affect monocyte chemoattractant protein 1 levels and antioxidant capacity of plasma. *The Journal of Nutritional Biochemistry, 20*(7), 521–529.
19. Zaid, H., Rayan, A., Said, O., & Saad, B. (2010). Cancer treatment by Greco-Arab and Islamic herbal medicine. *Open Nutraceuticals Journal, 3*, 203–212.
20. Zaid, H., Silbermann, M., Ben-Arye, E., & Saad, S. (2011). Greco-Arab and Islamic herbal-derived anticancer modalities: From tradition to molecular mechanisms. *Evidence-Based Complementary and Alternative Medicine, 2012.*
21. Saad, B., & Said, O. (2010). Chapter 10: Tradition and prospective of Greco-Arab and Islamic herbal medicine. In A. Dasgupta & C. Hammett-Stabler (Eds.), *Herbal remedies: Toxicity and effects on clinical laboratory test results*. Hoboken: Wiley.
22. Malik, M., Zhao, C., Schoene, N., Guisti, M. M., Moyer, M. P., & Magnuson, B. A. (2003). Anthocyanin-rich extract from Aronia meloncarpa E. induces a cell cycle block in colon cancer but not normal colonic cells. *Nutrition and Cancer, 46*(2), 186–196.
23. Kamei, H., Kojima, T., Hasegawa, M., Koide, T., Umeda, T., Yukawa, T., et al. (1995). Suppression of tumor cell growth by anthocyanins in vitro. *Cancer Investigation, 13*(6), 590–594.
24. Kamei, H., Hashimoto, Y., Koide, T., Kojima, T., & Hasegawa, M. (1998). Anti-tumor effect of methanol extracts from red and white wines. *Cancer Biotherapy and Radiopharmaceuticals, 13*(6), 447–452.
25. Zhao, L., Gu, J., Dong, A., Zhang, Y., Zhong, L., He, L., et al. (2005). Potent antitumor activity of oncolytic adenovirus expressing mda-7/IL-24 for colorectal cancer. *Human Gene Therapy, 16*(7), 845–858.
26. Zhang, X., Yeeleng, Y., Dong, W., Gu, C., & Feng, C. (2008). Novel omics technologies in nutrition research. *Biotechnology Advances, 26*(2), 169–176.
27. Jing, P., Bomser, J. A., Schwartz, S. J., He, J., Magnuson, B. A., & Giusti, M. M. (2008). Structure- function relationships of anthocyanins from various anthocyanin-rich extracts on the inhibition of colon cancer cell growth. *Journal of Agricultural and Food Chemistry, 56*(20), 9391–9398.
28. Yi, W., Fischer, J., & Akoh, C. C. (2005). Study of anticancer activities of muscadine grape phenolics in vitro. *Journal of Agricultural and Food Chemistry, 53*(22), 8804–8812.
29. Seeram, N. P., Adams, L. S., Henning, S. M., Niu, Y., Zhang, Y., Nair, M. G., et al. (2005). In vitro antiproliferative, apoptotic and antioxidant activities of punicalagin, ellagic acid and a total pomegranate tannin extract are enhanced in combination with other polyphenols as found in pomegranate juice. *Journal of Nutrition and Biochemistry, 16*(6), 360–367.
30. Lala, G., Malik, M., Zhao, C., He, J., Kwon, Y., Giusti, M. M., et al. (2006). Anthocyanin-rich extracts inhibit multiple biomarkers of colon cancer in rats. *Nutrition and Cancer, 54*(1), 84–93.
31. Cvorovic, J., Tramer, F., Granzotto, M., Candussio, L., Decorti, G., & Passamonti, S. (2010). Oxidative stress-based cytotoxicity of delphinidin and cyanidin in colon cancer cells. *Archives of Biochemistry and Biophysics, 501*(1), 151–157.
32. Kang, S. Y., Seeram, N. P., Nair, M. G., & Bourquin, L. D. (2003). Tart cherry anthocyanins inhibit tumor development in Apc sup Min/sup mice and reduce proliferation of human colon cancer cells. *Cancer Letters, 194*(1), 13–19.
33. Zu, X. Y., Zhang, Z. Y., Zhang, X. W., Yoshioka, M., Yang, Y. N., & Li, J. (2010). Anthocyanins extracted from Chinese blueberry (Vaccinium uliginosum L.) and its anticancer effects on DLD-1 and COLO205 cells. *Chinese Medical Journal, 123*(19), 2714.
34. Cai, H., Marczylo, T. H., Teller, N., Brown, K., Steward, W. P., Marko, D., et al. (2010). Anthocyanin-rich red grape extract impedes adenoma development in the ApcMin mouse: Pharmacodynamic changes and anthocyanin levels in the murine biophase. *European Journal of Cancer, 46*(4), 811–817.

35. Misikangas, M., Pajari, A. M., Päivärinta, E., Oikarinen, S. I., Rajakangas, J., Marttinen, M., et al. (2007). Three Nordic berries inhibit intestinal tumorigenesis in multiple intestinal neoplasia/+ mice by modulating β-catenin signaling in the tumor and transcription in the mucosa. *Journal of Nutrition, 137*(10), 2285–2290.
36. Singletary, K. W., Stansbury, M. J., Giusti, M., Van Breemen, R. B., Wallig, M., & Rimando, A. (2003). Inhibition of rat mammary tumorigenesis by concord grape juice constituents. *Journal of Agricultural and Food Chemistry, 51*(25), 7280–7286.
37. Stoner, G. D., Li-Shu, W., Zikri, N., Chen, T., Stephen, S. H., Chuanshu, H., et al. (2007). Cancer prevention with freeze-dried berries and berry components. *Seminars in Cancer Biology, 17*, 403–410.
38. Shin, D. Y., Lu, J. N., Kim, G. Y., Jung, J. M., Kang, H. S., Lee, W. S., et al. (2011). Anti-invasive activities of anthocyanins through modulation of tight junctions and suppression of matrix metalloproteinase activities in HCT-116 human colon carcinoma cells. *Oncology Reports, 25*(2), 567.
39. Srivastava, A., Akoh, C. C., Fischer, J., & Krewer, G. (2007). Effect of anthocyanin fractions from selected cultivars of Georgia-grown blueberries on apoptosis and phase II enzymes. *Journal of Agricultural and Food Chemistry, 55*(8), 3180–3185.
40. Zhang, Y., Vareed, S. K., & Nair, M. G. (2005). Human tumor cell growth inhibition by non-toxic anthocyanidins, the pigments in fruits and vegetables. *Life Sciences, 76*(13), 1465–1472.
41. Stoner, G. D., Kresty, L. A., Carlton, P. S., Siglin, J. C., & Morse, M. A. (1999). Isothiocyanates and freeze-dried strawberries as inhibitors of esophageal cancer. *Toxicological Sciences, 52*(suppl 1), 95–100.
42. He, J., Magnuson, B. A., & Giusti, M. M. (2005). Analysis of anthocyanins in rat intestinal contents impact of anthocyanin chemical structure on fecal excretion. *Journal of Agricultural and Food Chemistry, 53*(8), 2859–2866.
43. Carlton, P. S., Kresty, L. A., & Stoner, G. D. (2000). Failure of dietary lyophilized strawberries to inhibit 4-(methylnitrosamino)-1-(3-pyridyl)-1-butanone-and benzo [a] pyrene-induced lung tumorigenesis in strain A/J mice. *Cancer Letters, 159*(2), 113–117.
44. Ohara, A., & Matsuhisa, T. (2004). Effects of diet composition on mutagenic activity in urine. *BioFactors, 22*(1), 115–118.
45. Singletary, K. W., Jung, K.-J., & Giusti, M. (2007). Anthocyanin-rich grape extract blocks breast cell DNA damage. *Journal of Medicinal Food, 10*(2), 244–251.
46. Shih, P.-H., & Yen, G.-C. (2007). Differential expressions of antioxidant status in aging rats: The role of transcriptional factor Nrf2 and MAPK signaling pathway. *Biogerontology, 8*(2), 71–80.
47. Renis, M., Calandra, L., Scifo, C., Tomasello, B., Cardile, V., Vanella, L., et al. (2008). Response of cell cycle/stress-related protein expression and DNA damage upon treatment of $CaCo_2$ cells with anthocyanins. *British Journal of Nutrition, 100*(01), 27–35.
48. Bagchi, A., Papazoglu, C., Wu, Y., Capurso, D., Brodt, M., Francis, D., et al. (2007). CHD5 is a tumor suppressor at human 1p36i. *Cell, 128*(3), 459–475.
49. Boateng, J., Verghese, M., Shackelford, L., Walker, L. T., Khatiwada, J., Ogutu, S., et al. (2007). Selected fruits reduce azoxymethane (AOM)-induced aberrant crypt foci (ACF) in Fisher 344 male rats. *Food and Chemical Toxicology, 45*(5), 725–732.
50. Tramer, F., Moze, S., Ademosun, A. O., Passamonti, S., & Cvorovic, J. *Dietary anthocyanins: Impact on colorectal cancer and mechanisms of action*. In R. Ettarh (Ed.) Colorectal Cancer - From Prevention to Patient Care. InTech,Serbia.
51. Katsube, N., Iwashita, K., Tsushida, T., Yamaki, K., & Kobori, M. (2003). Induction of apoptosis in cancer cells by bilberry (*Vaccinium myrtillus*) and the anthocyanins. *Journal of Agricultural and Food Chemistry, 51*(1), 68–75.
52. Johnson, S. M., Wang, X., & Evers, B. M. (2011). Triptolide inhibits proliferation and migration of colon cancer cells by inhibition of cell cycle regulators and cytokine receptors. *Journal of Surgical Research, 168*(2), 197–205.
53. Thomasset, S., Teller, N., Cai, H., Marko, D., Berry, D. P., Steward, W. P., et al. (2009). Do anthocyanins and anthocyanidins, cancer chemopreventive pigments in the diet, merit development as potential drugs? *Cancer Chemotheraphy and Pharmacology, 64*(1), 201–211.

54. Rossi, M., Garavello, W., Talamini, R., Negri, E., Bosetti, C., Dal, M. L., et al. (2007). Flavonoids and the risk of oral and pharyngeal cancer: A case-control study from Italy. *Cancer Epidemiology, Biomarkers and Prevention, 16*(8), 1621–1625.
55. Bosetti, C., Gallus, S., & La Vecchia, C. (2006). Aspirin and cancer risk: An updated quantitative review to 2005. *Cancer Causes and Control, 17*(7), 871–888.
56. Stoner, G. D. (2009). Foodstuffs for preventing cancer: The preclinical and clinical development of berries. *Cancer Prevention Research, 2*(3), 187–194.
57. Huang, B., Zhao, J., Unkeless, J. C., Feng, Z. H., & Xiong, H. (2008). TLR signaling by tumor and immune cells: A double-edged sword. *Oncogene, 27*(2), 218–224.
58. Nagase, T., Ishikawa, K., Suyama, M., Kikuno, R., Hirosawa, M., & Miyajima, N. (1998). Prediction of the coding sequences of unidentified human genes. XII. The complete sequences of 100 new cDNA clones from brain which code for large proteins in vitro. *DNA Research, 5*(6), 355–364.
59. Krikorian, R., Eliassen, J. C., Boespflug, E. L., Nash, T. A., & Shidler, M. D. (2010). Improved cognitive-cerebral function in older adults with chromium supplementation. *Nutritional Neuroscience, 13*(3), 116–122.
60. Krikorian, R., Nash, T. A., Shidler, M. D., Shukitt-Hale, B., & Joseph, J. A. (2010). Concord grape juice supplementation improves memory function in older adults with mild cognitive impairment. *British Journal of Nutrition, 103*(05), 730–734.
61. Goyarzu, P., Malin, D. H., Lau, F. C., Taglialatela, G., Moon, W. D., Jennings, R., et al. (2004). Blueberry supplemented diet: Effects on object recognition memory and nuclear factor-kappa B levels in aged rats. *Nutritional Neuroscience, 7*(2), 75–83.
62. Shukitt-Hale, B., Lau, F. C., Carey, A. N., Galli, R. L., Spangler, E. L., Ingram, D. K., et al. (2008). Blueberry polyphenols attenuate kainic acid-induced decrements in cognition and alter inflammatory gene expression in rat hippocampus. *Nutritional Neuroscience, 11*(4), 172–182.
63. Williams, C. M., El Mohsen, M. A., Vauzour, D., Rendeiro, C., Butler, L. T., Ellis, J. A., et al. (2008). Blueberry-induced changes in spatial working memory correlate with changes in hippocampal CREB phosphorylation and brain-derived neurotrophic factor (BDNF) levels. *Free Radical Biology and Medicine, 45*(3), 295–305.
64. Spencer, J. P. (2010). The impact of fruit flavonoids on memory and cognition. *British Journal of Nutrition, 104*(S3), S40–S47.
65. Lau, F. C., Bielinski, D. F., & Joseph, J. A. (2007). Inhibitory effects of blueberry extract on the production of inflammatory mediators in lipopolysaccharide activated BV2 microglia. *Journal of Neuroscience Research, 85*(5), 1010–1017.
66. Galli, R. L., Bielinski, D. F., Szprengiel, A., Shukitt-Hale, B., & Joseph, J. A. (2006). Blueberry supplemented diet reverses age-related decline in hippocampal HSP70 neuroprotection. *Neurobiology of Aging, 27*(2), 344–350.
67. Youdim, K. A., Martin, A., & Joseph, J. A. (2000). Incorporation of the elderberry anthocyanins by endothelial cells increases protection against oxidative stress. *Free Radical Biology and Medicine, 29*(1), 51–60.
68. Ramirez, M. R., Izquierdo, I., do Carmo, B. R. M., Zuanazzi, J. A., Barros, D., & Henriques, A. T. (2005). Effect of lyophilised Vaccinium berries on memory, anxiety and locomotion in adult rats. *Pharmacological Research, 52*(6), 457–462.
69. Kang, T. H., Hur, J. Y., Kim, H. B., Ryu, J. H., & Kim, S. Y. (2006). Neuroprotective effects of the cyanidin-3-O-β-d-glucopyranoside isolated from mulberry fruit against cerebral ischemia. *Neuroscience Letters, 391*(3), 122–126.
70. Kruger, D. J., Greenberg, E., Murphy, J. B., DiFazio, L. A., & Youra, K. R. (2014). Local concentration of fast-food outlets is associated with poor nutrition and obesity. *American Journal of Health Promotion, 28*(5), 340–343.
71. Lacombe, A., Wu, V. C., Tyler, S., & Edwards, K. (2010). Antimicrobial action of the American cranberry constituents; phenolics, anthocyanins, and organic acids, against Escherichia coli O157: H7. *International Journal of Food Microbiology, 139*(1), 102–107.

72. Cisowska, A., Wojnicz, D., & Hendrich, A. B. (2011). Anthocyanins as antimicrobial agents of natural plant origin. *Natural Product Communications, 6*(1), 149–156.
73. Nohynek, L. J., Alakomi, H. L., Kähkönen, M. P., Heinonen, M., Helander, I. M., Oksman-Caldentey, K. M., et al. (2006). Berry phenolics: Antimicrobial properties and mechanisms of action against severe human pathogens. *Nutrition and Cancer, 54*(1), 18–32.
74. Burdulis, D., Sarkinas, A., Jasutiené, I., Stackevicené, E., Nikolajevas, L., & Janulis, V. (2008). Comparative study of anthocyanin composition, antimicrobial and antioxidant activity in bilberry (*Vaccinium myrtillus* L.) and blueberry (*Vaccinium corymbosum* L.) fruits. *Acta Poloniae Pharmaceutica, 66*(4), 399–408.
75. Cesoniene, L., Jasutiene, I., & Sarkinas, A. (2008). Phenolics and anthocyanins in berries of European cranberry and their antimicrobial activity. *Medicina, 45*(12), 992–999.
76. Wu, V. C.-H., Xujian, Q., Alfred, B., & Laura, H. (2008). Antibacterial effects of American cranberry (*Vaccinium macrocarpon*) concentrate on foodborne pathogens. *LWT--Food Science and Technology, 41*(10), 1834–1841.
77. Werlein, C. H.-D., Kütemeyer, G., Schatton, E. M., & Hubbermann, K. S. (2005). Influence of elderberry and blackcurrant concentrates on the growth of microorganisms. *Food Control, 16*(8), 729–733.
78. Chatterjee, A., Yasmin, T., Bagchi, D., & Stohs, S. J. (2004). Inhibition of Helicobacter pylori in vitro by various berry extracts, with enhanced susceptibility to clarithromycin. *Molecular and Cellular Biochemistry, 265*(1-2), 19–26.
79. Wang, S. Y., & Lin, H.-S. (2000). Antioxidant activity in fruits and leaves of blackberry, raspberry, and strawberry varies with cultivar and developmental stage. *Journal of Agricultural and Food Chemistry, 48*(2), 140–146.
80. Kong, J. M., Chia, L. S., Goh, N. K., Chia, T. F., & Brouillard, R. (2003). Analysis and biological activities of anthocyanins. *Phytochemistry, 64*(5), 923–933.
81. Sharma, R. (2001). Impact of solar UV-B on tropical ecosystems and agriculture. Case study: Effect of UV-B on rice. *Proceedings of Seawpit98 and Seawpit2000, 1*, 92–101.
82. Galvez, J., De La Cruz, J. P., Zarzuelo, A., Sanchez, D. L., & Cuesta, F. (1995). Flavonoid inhibition of enzymic and nonenzymic lipid peroxidation in rat liver differs from its influence on the glutathione-related enzymes. *Pharmacology, 51*(2), 127–133.
83. Varma, S. D., & Kinoshita, J. H. (1976). Inhibition of lens aldose reductase by flavonoids—Their possible role in the prevention of diabetic cataracts. *Biochemical Pharmacology, 25*(22), 2505–2513.
84. Beltoft, V. M., Binderup, M.-L., Frandsen, H. L., Lund, P., & Nørby, K. K. (2013). *EFSA cef panel (EFSA panel on food contact materials, enzymes, flavourings and processing aids), 2013. scientific opinion on flavouring group evaluation 21, revision 4 (fge. 21rev4)*. European Food Safety Authority.
85. Delgado-Vargas, F., Jiménez, A., & Paredes-López, O. (2000). Natural pigments: carotenoids, anthocyanins, and betalains—Characteristics, biosynthesis, processing, and stability. *Critical Reviews in Food Science and Nutrition, 40*(3), 173–289.
86. Zafra-Stone, S., Yasmin, T., Bagchi, M., Chatterjee, A., Vinson, J. A., & Bagchi, D. (2007). Berry anthocyanins as novel antioxidants in human health and disease prevention. *Molecular Nutrition and Food Research, 51*(6), 675–683.
87. Bridle, P., & Timberlake, C. (1997). Anthocyanins as natural food colours-selected aspects. *Food Chemistry, 58*(1), 103–109.
88. Oancea, S., & Oprean, L. (2011). Anthocyanin extracts in the perspective of health benefits and food applications. *Revista de Economia, 218*.

Printed in the United States
By Bookmasters